Clean Code in PHP

Expert tips and best practices to write beautiful, human-friendly, and maintainable PHP

Carsten Windler

Alexandre Daubois

BIRMINGHAM—MUMBAI

Clean Code in PHP

Copyright © 2022 Packt Publishing

Group Product Manager: Pavan Ramchandani
Publishing Product Manager: Elliot Dallow
Senior Editor: Keagan Carneiro
Senior Content Development Editor: Debolina Acharyya
Technical Editor: Joseph Aloocaran
Copy Editor: Safis Editing
Project Coordinator: Sonam Pandey
Proofreader: Safis Editing
Indexer: Tejal Daruwale Soni
Production Designer: Sinhayna Bais
Marketing Coordinators: Anamika Singh and Marylou De Mello

First published: October 2022

Production reference: 1061022

Published by Packt Publishing Ltd.
Livery Place
35 Livery Street
Birmingham
B3 2PB, UK.

ISBN 978-1-80461-387-0

www.packt.com

To Laura and Britta. Without you, I'm nothing.

– Carsten Windler

To Pierre-Louis, François, and Xavier, who introduced me to the world that I wrote a book about.

– Alexandre Daubois

Contributors

About the authors

Carsten Windler started working with PHP in version 3, and it has been his main programming language since then. Over the years, he also headed multiple development teams and departments, which has provided him with a sound understanding of the difficulties of writing high-quality PHP code in a team.

He currently works as lead backend engineer at Plan A in Berlin, where he combines his passion for writing code with his wish to combat climate change.

I would like to thank the entire PHP community for their non-stop efforts, especially those who write great open source software in their spare time, making life easier for countless developers.

Alexandre Daubois started learning PHP at the same time he was taught about Symfony. At first mainly interested in more low-level programming languages such as ASM, C, and C++, he's since become involved in the PHP and Symfony ecosystem more and more. After giving conferences at official Symfony events, contributing to the code source of the framework and its documentation, and working at SensioLabs (the company that created Symfony), writing a book about PHP was logically the next step.

About the reviewers

Brad Orlemanski graduated from Appalachian State University with a bachelor's degree in science. For 22 years, he's been working in web development, with 9 years of experience using PHP. When not working on web development, Brad likes to work on other projects, such as building IoT devices using Arduinos and Raspberry Pis and developing educational games with Unity3D and C#. Brad is excited about the upcoming launch of his own start-up company.

Hossein Rafiei is a backend and frontend developer with 13 years of experience in implementing and developing user interfaces, and developing new RESTful and GraphQL APIs. Also, he has advanced knowledge in testing, debugging, and training staff in frontend and backend technologies.

He is continually evaluating and upgrading his skills to stay at the cutting edge of backend and frontend development. He has a strong background in project management and customer relations. Also, he is committed to having an impact on the future of any company that he works for and possesses the enthusiasm and commitment to learn and develop his career.

His favorite tech company is Amazon, and he is currently working on AWS and microservice technologies with a dream to build better web and mobile applications for society.

Marco Dal Monte has a master's degree in computer science from the University of Padova, Italy, where he specialized his studies in artificial intelligence.

After university, he worked for a brief period in Italy before moving to London.

He works mainly in PHP, using Symfony and Laravel as frameworks, and he is interested in software engineering methodologies, DevOps, and system administration in Linux.

He can confirm that this book is going to help many developers to improve their code, as he is using these methodologies and tools in his daily work with great results.

Table of Contents

4

It is about More Than Just Code   33

5

Optimizing Your Time and Separating Responsibilities   43

6

PHP is Evolving – Deprecations and Revolutions   51

Part 2 – Maintaining Code Quality   61

7

8

9

Preface

PHP has evolved over several decades from a simple scripting language for creating HTML pages to a feature-rich language with an extensive ecosystem. Because the vast majority of websites are still powered by PHP, it is one of the cornerstones of the internet.

While still beginner-friendly, it can be used to implement everything from small websites to enterprise applications used worldwide. However, the low entry barriers of PHP sometimes lead to code that is difficult to understand and impossible to maintain.

With this book, we want to introduce you to the world of **Clean Code**. You will learn a lot about the theory and also, how to apply the knowledge you learned in the real world. You will learn which tools will support you on this journey and which best practices you should use to be able to implement Clean Code successfully in your team.

Who this book is for

This book is aimed at early career PHP developers who want to understand the foundations of high-quality PHP code, and seasoned PHP developers who wish to update themselves on the latest best practices.

What this book covers

Chapter 1, *What Is Clean Code and Why Should You Care?*, introduces the main subject of the book.

Chapter 2, *Who Gets to Decide What "Good Practices" Are?*, explains how those "rules" are decided.

Chapter 3, *Code, Don't Do Stunts*, shows why you should consider being pragmatic rather than trying to show off skills.

Chapter 4, *It Is About More Than Just Code*, explains why a clean code perimeter is greater than just writing source code.

Chapter 5, *Optimizing Your Time and Separating Responsibilities*, explains how to become more productive by creating new habits.

Chapter 6, *PHP Is Evolving – Deprecations and Revolutions*, provides a quick overview of the most awaited features introduced in PHP, helping to write clean code.

Chapter 7, *Code Quality Tools*, teaches you about tools that will help you write clean, maintainable code.

Chapter 8, *Code Quality Metrics*, looks at all the metrics you need to know to assess your code quality.

Chapter 9, *Organizing PHP Quality Tools*, shows you how to keep your tools organized.

Chapter 10, *Automated Testing*, introduces you to automated testing and explains why you should do it.

Chapter 11, *Continuous Integration*, explores how to maintain code quality consistently and over time.

Chapter 12, *Working in a Team*, introduces you to the best practices for working in a team of developers.

Chapter 13, *Creating Effective Documentation*, demonstrates how to create useful and living documentation.

To get the most out of this book

You will need PHP 8.0 or above installed on your computer. All code samples have been tested using PHP 8.1 on Linux and macOS. With little adjustments, they should work on Windows too, depending on your setup.

Software/hardware covered in the book	Operating system requirements
PHP 8.0 and above	Linux, macOS, or Windows

To follow all chapters, you will also need to install Composer. You will find more information about it in *Chapter 9*, *Organizing PHP Quality Tools*, or at `https://getcomposer.org`.

If you are using the digital version of this book, we advise you to type the code yourself or access the code from the book's GitHub repository (a link is available in the next section). Doing so will help you avoid any potential errors related to the copying and pasting of code.

Download the example code files

You can download the example code files for this book from GitHub at `https://github.com/PacktPublishing/Clean-Code-in-PHP`. If there's an update to the code, it will be updated in the GitHub repository.

We also have other code bundles from our rich catalog of books and videos available at `https://github.com/PacktPublishing/`. Check them out!

Download the color images

We also provide a PDF file that has color images of the screenshots and diagrams used in this book. You can download it here: `https://packt.link/b08Jl`.

Conventions used

There are a number of text conventions used throughout this book.

`Code in text`: Indicates code words in text, database table names, folder names, filenames, file extensions, pathnames, dummy URLs, user input, and Twitter handles. Here is an example: "So, we will naturally create a service like `UserRemover`, which will perform these two tasks in a row."

A block of code is set as follows:

```php
<?php
class Example
{
    public function doSomething() bool
    {
        return true;
    }
}
```

When we wish to draw your attention to a particular part of a code block, the relevant lines or items are set in bold:

```
{
    ...
    "scripts": {
        "analyse": [
            "tools/vendor/bin/php-cs-fixer fix src",
            "tools/vendor/bin/phpstan analyse --level 1 src"
        ],
        "post-update-cmd": "composer update -d tools",
        "post-install-cmd": "composer update -d tools"
    }
}
```

Any command-line input or output is written as follows:

```
$ php phploc src
```

Bold: Indicates a new term, an important word, or words that you see onscreen. For instance, words in menus or dialog boxes appear in **bold**. Here is an example: "Hovering the mouse pointer over **TestClass** will show a popup window with an explanation saying **Undefined type TestClass**."

> **Tips or important notes**
> Appear like this.

Get in touch

Feedback from our readers is always welcome.

General feedback: If you have questions about any aspect of this book, email us at customercare@ packtpub.com and mention the book title in the subject of your message.

Errata: Although we have taken every care to ensure the accuracy of our content, mistakes do happen. If you have found a mistake in this book, we would be grateful if you would report this to us. Please visit www.packtpub.com/support/errata and fill in the form.

Piracy: If you come across any illegal copies of our works in any form on the internet, we would be grateful if you would provide us with the location address or website name. Please contact us at copyright@packt.com with a link to the material.

If you are interested in becoming an author: If there is a topic that you have expertise in and you are interested in either writing or contributing to a book, please visit authors.packtpub.com.

Share Your Thoughts

Once you've read, we'd love to hear your thoughts! Scan the QR code below to go straight to the Amazon review page for this book and share your feedback.

https://packt.link/r/1804613878

Your review is important to us and the tech community and will help us make sure we're delivering excellent quality content.

Part 1 – Introducing Clean Code

The purpose of this first part is to introduce the concepts of clean code and the theory behind it. This part is intended to be quite theoretical, although concrete examples will be discussed as the chapters progress. The first approach of clean code is proposed in order to explain its usefulness and why it is necessary in the life of a developer.

This section comprises the following chapters:

1

What Is Clean Code and Why Should You Care?

If you are new to **PHP: Hypertext Preprocessor** (**PHP**) and software development in general, there's a good chance you haven't come across clean code yet. We all began here; everybody must be aware of this. Clean code is more than just a set of rules. It is a real mindset—a way of thinking about what you are creating.

We often come across developers that have been in **information technology** (**IT**) for many years, even decades, and never had to worry about clean code. This may be for several reasons for this: working alone on a project for many years or developing on a legacy code base are a few of many reasons "clean code" did not reach their ears or wasn't one of their concerns.

Become aware of clean code—the earlier the better. As with many other habits, once you have adopted a way of working with code, it is difficult to change this. It may be because you are not convinced at all about what clean code is about, you are not sure it applies to you, or you may just be lazy about applying new "rules" because yours are already working for you (be assured that everyone's been through this). But if you care about it and you are taking note—well, you've already done the hardest part, because this means that you are ready to change your habits. Don't misunderstand us—the point is not to change everything you've learned up to now; it's more a case of improving the skills you already have.

You may not be aware of this, but it is pretty certain that you already have some habits that totally suit clean-code principles and that are natural to you and completely new to other developers. You should note that the goal of clean-code principles is not to be controversial at all.

The main topics we'll cover in this chapter are listed here:

- What this book will cover
- Understanding what clean code is
- The importance of clean code in teams
- The importance of clean code in personal projects

What this book will cover

You may have already heard about software craftsmanship books, clean-code books, and so on. What is the difference between this book and other ones? Well, we realized when we are a neophyte in development and discovering new programming languages, or even getting into programming and source code for the first time, it may be hard to understand how to apply these principles you read about everywhere. Programming languages do not all offer the same features—some are evolving fast, while some are not evolving anymore. Notions and the way things are named in one language may not be valid in other languages. That is why this book focuses on clean code in PHP especially.

You will not have to think about how to implement these principles and rules in PHP, which is something less to remember when absorbing knowledge. You will find examples and tools applying directly to you, used in professional projects, and proven by the industry. This way, we can get straight to the important points we would like you to understand. You can put your knowledge into practice immediately while reading this book and as a result, learn faster. Additionally, you do not need 10+ years' experience to understand what will be shown. Only basic knowledge of PHP is required to fully understand what will be exposed.

This book is a compilation of years of experience—years of practice in the field, faced with real problems that had to be solved according to technical constraints, functional constraints, and time and money constraints. This will not be yet another book full of utopian principles that cannot be applied in real life.

More than fitting directly to your environment by talking about PHP only, this book can be read in the order you want. It is divided into two distinct parts. The first one (which you are currently reading) will expose a bit of theory about what is clean code and what are basic principles of it, directly applied to the PHP language up to version 8.2, and so on. The second part will be focused on practical tools you can use to ensure you are following the right rules the correct way, setting up an environment and your **integrated development environment** (IDE) in order to be as efficient and clean as possible, and getting metrics on your code, automated testing, writing documentation, and more. This means you can skip parts, take them in the order you want, and learn at the speed you want.

You will not have to read the whole theoretical part before getting into practice—you can dive into tools and concrete examples right now. You can also focus on the first part before getting into the practical aspects in order to be sure you fully understand what will be explained later. It is up to you.

Understanding what clean code is

Clean code is the act of writing code by thinking of the future—that is, thinking of the people who will collaborate with you or on the project without you. And when we are talking about "the people who will collaborate with you", this may even be yourself. If you never did, you should try to read and maintain the source code you wrote a few months (or years) ago. Isn't it challenging? Well, do not worry—it is hard for everyone. You will surely find this complicated for many years, and there is no quick fix to this. It is *the same process for everyone.*

There are many reasons for this, as outlined here:

- You are still learning new ways to code every day

- You are practicing and you're getting better at coding

- Your way of thinking may be evolving because of the people you meet or simply because you are getting older

- Innovative technologies and libraries are released every day, and some ways of doing precise tasks are getting deprecated

- The language you are working with (here, PHP) is evolving and new standards are appearing

Talking about PHP, it is remarkably interesting to see the language is evolving greatly and quickly these last few years. Many called for the death of PHP a few years ago. It is true that before PHP 7, the language was truly not going well. PHP 7 and newer versions are a huge breath of pure air for us. New versions are being released frequently and features are being added to PHP (just look at the `match` expression and the support of enumerations in PHP 8.1!). Strict typing has also been introduced, which is a gigantic step forward too. We will dive into PHP evolutions that ease the writing of clean code later.

It does not really matter which programming language you are using—undeniably, you will get scared each time you look at some old code you did yourself. Writing your code in a "clean" way will help you to minimize this effect and understand it better, instantly, years later. By doing this, you are giving yourself a treat, because you will be coding in a way you are more and more used to—a way of writing many people are used to.

It may be the most impressive thing about clean code: it does not matter which technology you are using, which library, or even which language. It does not depend on the country you are living in and it does not depend on your years of experience, or even if you are coding for a living or as a hobby without financial intentions behind it. Clean code is a skill that at some point, every developer will improve at and face in their life. Of course, you will get better at it if you focus on it, but just learning from your experience of what works and what does not in an IT project is already, in a way, learning clean code. More than the rules, tools, and tips you will find in this book, clean code is a question of experience and years of practice. During your developer career, you will meet different challenges and diverse ways of thinking by meeting people and collaborating with numerous different teams.

You should always remember that coding is a team job. It is rare now to work alone on a project without the intervention of other developers—if it still happens at all. It may be a professional project, an open source project, or even a private personal project. Saying that other developers will help with your project does not necessarily mean that they will write some code. They can also try to find solutions with you to the technical challenges you will be facing, and that applies to every type of project.

In any case, doing things properly will ease a crucial step we all go through when getting into a new project: the onboarding process.

The importance of clean code in teams

The onboarding process is never easy. You are getting into a new project, and many people around you are already familiar with it. You must learn everything: who does what in the team, what the exact requirements are, the technical challenges the team has been through, and what is coming next. This process can last several weeks, and even several months in some cases (it can also last for more than a year on some gigantic legacy projects, but that is another subject). If you can save yourself the process of learning the coding conventions because you already know them thanks to clean-code principles, then it is one thing you will not have to think about and you can focus on other things.

Being able to write clean code is like speaking fluently in the same language as the other people in the team: it makes it easier to communicate and write code in the same way without noticing who has written which part.

Clean code will help you with code longevity. When working as a team on a project, there are great chances the source code will be maintained for at least a few years. The code will continue to be maintained years after your departure. It is important to do things right because you will not be here forever to explain everything you did to the person still working on the project.

Writing code cleanly will avoid the apparition of what we call **single points of failure** (**SPOFs**). These SPOFs are the worst nightmare for any project. They are also symptomatic of legacy projects. They can be described as an entity (a developer, a technology, a library, and so on) that holds the whole project by itself. If this entity fails, everything crumbles.

A concrete example of an entity failing is the departure of a developer. If they are the only person to utterly understand how the project works and is considered the "superhero" of the project, then it is a big problem. This means the project will not be able to continue without them, and it will be an absolute pain to maintain the project after their departure.

One excellent example of SPoF avoidance was given by Fabien Potencier, the creator of the Symfony framework, during the *SymfonyWorld Online 2021 Winter Edition* convention. Even if Symfony is an open source project, he has been the person working the most on it since its creation. He explained during a conference that one of his main tasks now is to transmit the maximum amount of knowledge about the framework to the maximum number of people so that Symfony does not rely on him exclusively anymore. He absolutely wants to avoid being the SPOF of Symfony, and naturally wants Symfony to be maintained and developed, even after he's no longer working on it himself anymore.

One of the numerous ways to avoid this problem is by writing clear, concise, and understandable code—writing code in a way everybody can easily understand what is happening. After getting into this book, you may have a better idea of how to achieve this.

Talking about teamwork, another critical point is code reviews. Even if we will be getting into the details of these later in this book, let's talk about them quickly here. If you have never heard of a code review, it is simply a process where other developers of the projects are reviewing, reading, and commenting on the changes you want to bring to the code base. This process is mandatory in most open source projects and *highly* recommended for any project, given the benefits it brings. You can now easily imagine that if everyone has their own way to write code, those reviews are going to take way longer. This goes from the way you format your code or how you separate files, classes, and so on. If we all speak the same language, then we can focus on things that really matter during the code review: if the functional need is respected, whether there are any bugs, and so on.

We will see in the next chapters that clean-code definition often depends on the team you are working with. Even if there are some general rules, it mostly depends on what the developers you are working with are used to. Again, the reason is habits but also consistency. This way, the team can work faster and better together.

If you happen to work in an area where several teams are gathered in the same place, you will inevitably notice that each team has its own set of rules that works well for them, but also common rules that will allow you to talk with everyone in the same language. Because solutions to your problems are not always found in your direct team but sometimes also from people surrounding you without working with you on the same project directly, it will be so much easier to help each other if basic rules are shared between groups of people.

The importance of clean code in personal projects

You may think that clean code is less important for personal projects, even just a tiny bit. You would be wrong if you think like this for many reasons, as set out here:

- How can you be sure that nobody will ever be involved in the development in the future?
- If you make your project open source, don't you want the world to see that you are coding cleanly and be proud of this?
- You will probably want to improve your project repeatedly. If you write bad foundations, it will be a nightmare to maintain and add new things without the fear of breaking anything. Sometimes, it is impossible to add new features because of bad code writing. In the worst cases, you will have to entirely rewrite your application. You do not want to do that.

As mentioned earlier, try to read the code you wrote a few years ago. There's a great chance you will not be able to understand briefly what you wanted to do at that time. You may think the comments you put (if you do this) are here to help, but let's ask two questions, as follows:

- Who truly reads code comments?
- What about writing code in a way that comments are unnecessary, and it is so clearly written it can be read like a book?

This is what it is all about: being able to read the source code like simple sentences without having to stop on a line to understand it. Also, to repeat what was said about clean code in teams: just because you are creating a project on your own does not mean that you won't need external help. In fact, it is almost a certainty that one day, you will need the help of someone for something. Again, if you share the same set of common rules, it will be much easier for the helper to bring in their ideas. Even in personal moments and projects, a developer's job is teamwork. Always.

Clean code is a mindset, and this mindset includes being proud of what we created—the taste of a job well done. It will maybe take a few years to fully understand what it is all about, and we always have new things to learn and situations we are facing for the first time. Clean code is to move toward the moment when you look at your code and tell yourself you are completely relaxed about showing your code to anybody.

As the saying goes, "write code as if the next developer to run it knows your address". And this should also apply to personal projects because you are that very next developer.

Summary

Are we done yet with the theory? Well, kind of. We defined together what clean code consists of. We gave a mutual definition of clean code. By having the same definition of clean code, we are one step deeper into it, and ready to dive into more advanced principles in the next section.

But of course, we are not done yet. Even if you agree with the definition we presented, you may of course have a lot of questions already. And the most important questions should be the following: Who decides these rules for you? Who has the power to impose this vision? This is what we are going to see together in the next chapter.

2
Who Gets to Decide What "Good Practices" Are?

Good practices are great but knowing who decides them and where they come from is better. It is no secret that when you fully understand what you are doing, you immediately feel better and more comfortable. The same thing applies to good practices. Why should you believe without question these precepts decided by people you do not know and who have never collaborated with you on your project?

You could say that the people dictating these principles have more experience than you and know this world better than you do. Two things. First, maybe one day you will have more experience than they do. Maybe you will be better. Maybe you already are. Second, years of experience are not everything. It is common to see developers with 20 or 30 years of experience who are completely out of date or with habits from the last century. Years of experience can be an argument, but not the only one. Computing evolves at an exceptional speed, and the web world is even more affected by this.

We're going to see together in this chapter where best practices have originated from: were they really invented and decided by a precise group of people? What are the different existing clean-code principles that you can already apply to your projects starting from now? Your way of thinking might change once you know them.

Here are the main topics we'll cover in this chapter:

- Who decides these things anyway?
- Best practices—where do they really come from?
- Being context-aware
- Being consistent—get results quicker

Who decides these things anyway?

One thing we are going to see is that you should always question "good practices" and never consider them as a general truth that you must respect without understanding why. One excellent way to improve yourself is to ask every time you do not agree with someone's review or point of view. Developers are fascinating as they can find unlimited ways to solve a single problem—unlimited solutions for the same result. Even if this can be seen as something a bit tiring sometimes, it is always interesting to understand why a developer wants to solve a problem in a different way than you. There are multiple goals to this, as follows:

- **You will improve your communication skills**: If you want to communicate and be understood, you will have to explain your problem clearly.

- **You may learn new ways of doing things**: We are all using the same language, but we all have different experiences with it. These different career and life paths can bring terrific ideas to the table.

- You are reinforcing your relationship with this developer, which will make it easier to discuss further subjects in the future.

- You are improving your teamwork skills and all the participants in the conversation are improving their skills, respectively.

That is a pretty long list. A lot of soft skills are improved just by discussing with other developers a way to solve a given problem. Being able to explain a situation clearly is way harder than it seems.

And this is also why you should always ask for further explanation when you disagree with someone. First, there can be a misunderstanding about the exposed problem. This is a common situation where multiple parties are disagreeing: the problem was not clear at first. Each party is trying to justify what they understood. You can easily imagine the mess it brings.

Being a good developer is (also) being able to justify and explain all your choices. No more: "We always made things like this; there are no other reasons to do it this way." When you are deciding or telling someone to follow some guidelines, you must always be able to justify and explain clearly why your way is the best for you. It may not be the best way to do it objectively, but if you are able to explain why it is best suited to you, it will make you look way more open-minded.

That being said, you may now understand where we are going with this: no one holds the absolute truth. If someone is confident enough to say so, you should always be careful.

Best practices – where do they really come from?

When we are talking about "best practices," we can differentiate three cases, as follows:

- **Principles that have been proven for decades to work, which also are deduced from common sense**: In this category, we can—for example—find design patterns. In short and if you do not know them, these are tools that fix recurrent programming problems. They have been here for decades and are known by millions of developers.

- **Choices made because we had to make them**: Here, we can find things such as code style, naming conventions, and so on. Technically, it does not matter if you would like to use *camelCase* or *snake_case* to name your files. But if everyone is following the same rule, it is easier for everybody to understand each other.

- **"Technical" best practices**: Some "best practices" are actually dictated by technical constraints and features. A concrete example is "method name guessing" in **PHP: Hypertext Preprocessor** (**PHP**). Let's say you would like to guess "getters" (accessors) and "setters" (mutators) of an attribute of a PHP class. You are likely to try methods starting with `get`, `set`, `is`, and so on. If everyone has their own rules about naming accessors and mutators, you can be sure that someday and without warning, things will blow out.

Design pattern principles

The design patterns case describes objective solutions to solve problems. You can dislike them and how the code is being organized by them but cannot say that they are objectively bad. Because your thoughts on them do not matter, they work.

Talking about principles that have been present for decades, we can highlight four famous ones: **DRY**, **KISS**, **YAGNI**, and **SOLID**.

DRY

DRY stands for **Don't Repeat Yourself**. This principle simply states that you should never have, in your application, two authorities doing the very same thing. It may sound obvious, but applying this may not always be a reflex, especially when you are new to application programming. Having the same responsibility at two separate places in your code means maintaining those two places every time you fix something. It means having to think about these two places at each change (and someday, you are going to forget one, for sure). Also, how is any developer maintaining your source code supposed to know which one to use if two things have the same responsibility?

KISS

KISS stands for **Keep It Simple, Stupid**. Sometimes, we complicate our lives. We can see two main reasons for this, as follows:

- First, we try to do complicated things with our code, but these stunts do not bring anything valuable and complicate the code. We will see later in this book in detail why we absolutely need to avoid this.

- The second reason is a lack of perspective on what we are doing. We have spent many hours trying to solve something, and we are too much "into it." Some rest is necessary to get this perspective and, sometimes, start it all over again.

Both cases are recurrent and prevent us from going straight to the point and keeping things simple. When you feel you are going "too far," think of this acronym to get back on the rails.

YAGNI

YAGNI stands for **You Aren't Gonna Need It**. In some way, it goes hand in hand with the KISS principle. It is quite common (not to say part of our daily lives as developers) to want to find a solution to a problem by thinking about the future. Indeed, we regularly have this thought: "If tomorrow I need to do such and such a thing, at least this will already be in place." The reality is that, in general, no—we will never have the need that we want to try to foresee. Then, by trying to get ahead of a task that may never exist and whose functional constraints are unknown, not only do we waste time, but we also complicate our lives by thinking too far ahead. We move away from our initial goal, which is to find a quick, viable, and robust solution.

We are not psychic, and we cannot know all the problems that will appear if the task we predicted appears. You will quickly realize that if you keep things simple and without superfluous additions to try to get ahead of the game for the next day, you will have a healthy, no-frills code base. This means a faster understanding of the code, greater ease in navigating through it, and making changes when they are really needed. Plus, you will probably save yourself a lot of bugs. It is always complicated (if not impossible) to justify that a bug was made in your source code because you developed and spent time on something that was not asked of you. If your job as a developer requires you to collaborate directly with the customer, you should know that the customer will not pay you for something they did not ask for. You will have worked for free, which is never ideal.

Caution: It is obviously necessary to take this on a per-case basis. We can take as an example *magic numbers*. Magic numbers are constant values, mainly numbers, hardcoded and without any explanation of their meaning.

Verdict: Two weeks later, everyone has forgotten what this number corresponds to. We will then think of using well-named code constants. However, there is a huge chance that the value of this code constant will never change because the needs would have changed. At first sight, it would be strange to want to declare a constant to use it everywhere. The purpose of code constants is to add semantics to fixed values and allow us to easily change that value everywhere it is used in the code, at once. This is, in a sense, contrary to YAGNI (since these values will probably never change).

However, we can see the value of using constants. Perspective and reflection are always necessary, whatever the clean-code principles applied.

SOLID

Finally, maybe the most famous one, **SOLID**. Let's see what these letters stand for:

- S stands for **single-responsibility principle** (often abbreviated to **SRP**). Very simply, it means that a class in your code must respond to only one task. Obviously, the size of that task is the key point here. We are not talking about creating a class with only one available method. Rather, we are talking about creating a logical breakdown. A very concrete example is the **model-view-controller** (**MVC**) architecture. The important thing to remember is that you must avoid having catch-all classes, grouping together database operations, **Hypertext Markup Language** (**HTML**) rendering, business logic, and so on. The breakdown must be logical. An example of a breakdown could be a class for generating HTML, a class for database interactions for a given object, and so on.

- O stands for **open/closed principle** (**OCP**). You will often find the following definition for this principle: a class must be open to extension and closed to modification. Concretely, in the code, this is materialized by the strong use of polymorphism and the use of interfaces rather than by conditional branching with multiple `if` and `else` statements. Indeed, if you use conditional branching, you incur a modification of the class, and this can quickly become unmanageable if you have more than two cases. By extending a class and overloading the methods you are interested in, you get concise code, well broken up and without branching of several hundred lines.

- L stands for **Liskov substitution principle** (**LSP**). Behind this complicated name is actually something quite trivial. In fact, this principle simply says that when you use an interface implementation, you should be able to replace it with another implementation without having to modify the implementation in any way. In the code, this translates into the fact that the implementations of an interface must be similar, especially regarding the return values of the methods (if one implementation returns a string when calling the `foo` method and another implementation returns an object, it will be complicated). Fortunately, the typing of return values exists in recent versions of PHP, limiting the possibilities of violation of this principle.

- **I** stands for **interface segregation principle (ISP)**. This describes that a class implementing an interface should not be forced to implement or depend on other methods of the interface that it does not use. Concretely, this principle will prevent you from creating an interface with dozens of methods in it that are there *just in case*. It is better to create several interfaces (joining the principle of single responsibility), even little ones, with a very precise goal and responsibility. Then, you'll be able to implement them unitarily in your class.PHP allows a class to implement as many interfaces as we want, so this works perfectly fine. Thanks to this, the implementation we describe will only know the methods that are useful to it. Otherwise, we would end up with dozens of methods with an empty body or returning a `null` value. When you put it like that, you realize that it does not sound like very "clean code."

- Finally, **D** stands for **dependency inversion principle (DIP)**. Once again, this is a term that may seem complex at first glance but hides a much simpler reality. Very concretely and in the code, it materializes by using interfaces and abstraction rather than implementations. For example, when you type the argument of a method, you use the interface as the type. This will allow you to make full use of polymorphism and to take full advantage of Liskov substitution. By using the interface as a type, you will be able to send as an argument to the function any implementation of the interface. To give an example, if you need to send emails in an application, chances are you will create a `MailerInterface` interface. You will then have one implementation per mail service. By typing the argument with the interface, the method will be able to receive any implementation and use the right email service for your case.

We can quickly realize that these principles are very much linked. They work together, and they allow a strong decoupling, a strong separation of responsibilities, and a fluid thinking when you have these principles in mind while writing code. It can be extremely helpful to remember these principles; at least, it is a particularly good thing to know their existence. Apart from the SOLID principle, you can see that KISS, DRY, and YAGNI are pretty common-sense and logical. Remembering them from time to time can be beneficial and can help us to put up barriers when we get a little off track.

Bonus – Scouts' principle

Something that can also be added, which is also a common-sense principle, is the "Scouts' principle." We all know those groups of young people and teenagers that act with a benevolent purpose and show a great deal of altruism. The Scouts go camping in the woods, build a fire, and spend the night there. Once they get up in the morning, they might put out the fire and put away their stuff, but most of all, they clean up the place to make it cleaner than it was before they arrived (at least, in theory).

As a developer, it is the same. Be a Scout. When exploring and browsing the code, if time and context allow, it is often a particularly clever idea to clean up where you see some technical debt. If you are going through places in the code base and think "this is really bad," maybe this is an opportunity to make it more manageable and cleaner. If everyone gets on board, the quality of the project's source code can rise very quickly.

Of course, this "Scouts' principle" must be done in accordance with your project constraints, time constraints, and customer needs. Moreover, it is quite risky, and you must know when to stop. When you send your changes to your team for review, the changes and cleanup you have done must be consistent. You do not want to rewrite half the application every time you find a little thing, which leads to another, then another, and so on. It's more about cleaning up the things that are relevant to what you're doing. It can be extremely complicated to stay focused and fixed on your context. There is no genuine answer to "when to stop"; it will depend a lot on the time you have and your task. However, there is nothing to stop you from writing down things that you want to come back to but that unfortunately were not related to what you were doing, that seemed too energy- and time-consuming, or that simply require further reflection with the team.

On the opposite side, code style, naming conventions, and similar things are subject to tastes and habits. Everyone has their own tastes and habits, so decisions need to be made. As we discussed earlier, it is so much easier to talk together when we all follow the same rules.

So, who decides these best practices in a team or an organization? Well, generally, it is a consensus after lengthy discussions about the subject at the beginning of a project. Because yes—"best practices" are not something that you can always apply everywhere, and you should be aware of this. You should be aware of the context you are in.

Being context-aware

Here, we enter one of the most important parts when we talk about clean code. If there were only one thing to remember, it would be this. We may regularly talk about the rules defined by other developers, object principles, and the principles of clean code, but nothing will ever be as good as what we are going to talk about here: it is about being aware of your context. One thing that is missing from many books and articles about clean code is the feeling that it is relevant to everyday life. A developer's life is made up of unexpected events, technical constraints, impossibilities to do some things, or being forced to do some other things.

There are as many ways of doing things as there are projects. Each project has its own history, technical decisions, and constraints. As a result, we end up with many theoretical principles that are not applicable or that would break the coherence of the project. Good practices may dictate how you name variables, how you name your classes and methods, how you name your files, or how you make up the tree structure of your project. However, what if this goes against what has been set up in the project? This is a problem that arises very regularly, especially on so-called "legacy" projects.

The answer is both simple and complicated. It is simple because it can be summed up in one sentence: discuss it with your team and the other people working on the project. This is where the answer can become complicated because it will most likely start debates (sometimes heated) within the team, and this is quite normal. But the important thing is this: you must not only find a common ground that everyone will respect to be consistent with the project, but also find ways that work best for you and your team.

This may be the cornerstone of all these discussions: you may have good practices and clean-code principles, but maybe some rules are worth bending because it just works better for you and your team. This is a perfectly valid reason to deviate from certain principles (not all of them, of course, otherwise it is total anarchy). Sometimes, no common ground will be found, and that's when best practices and other principles can prevail in dictating a rule that everyone will have to follow, if not one that works for everyone.

And that is when we come back to a point made earlier. Having common rules in a team of developers helps to make it easier to be understood. Navigating through the project is easier, as is navigating through files. By all speaking the same language, we understand each other better. Whether it is helping someone or being helped, it is always the same story. If common naming and indentation rules are respected, not only do you avoid sterile debates that would distract you from your initial goal, but if you feel that the code written by your teammate is written by you, it is a win-win.

We can summarize the choice of deciding on good practice and "classic" principles of clean code in this way:

Has your situation already occurred in the past in the project?

- *Yes*. Does it respect the principles of clean code and the good practices dictated by the tools you use (such as *Symfony*'s good practices)?

  - *Yes*. In this case, you just have to follow the way the situation has been handled in the past.

  - *No*. You should talk to your team to find out why. There may be historical reasons due to functional and/or technical constraints, or maybe there is no reason at all. In this case, if you agree, you can follow the good practices (and even update the other parts of the code concerned if you have the possibility, and the time, to respect the "Scouts' principle" we discussed earlier, in the previous section).

- *No*. Is it possible to apply good practices and clean-code principles?

  - *Yes*. Perfect! All you have to do is apply them to the best of your ability, in accordance with the other practices in place within the team and the project.

  - *No*. In this case, you should discuss it with your team, but perhaps also discuss it with people outside the project who have had this case. Again, these discussions can take quite a long time, and an answer will not always be found immediately. Debates will be raised, for the better. Once you have found a consensus or discussed it at length, you can either rethink the application of clean-code principles and good practices or apply rules that have been set up with the team for this kind of situation.

Here, we see very clearly, once again, that the key to success will be communication and debate. Everyone has their own point of view and approach to the problem, so it is never all black and white. But keep one thing in mind: you should avoid making decisions alone about a practice that could be debated. Avoiding making a decision alone does not mean you should not make one. Rather, what we mean by this is to think through the options, weighing up the pros and cons of each. And again, *be able to justify each proposal* you offer to the other people working on the project. You can be sure that you will be highly appreciated by your team if you offer several solutions to the problems with a good justification of why these solutions are appropriate, but also the risks that these solutions entail. You will quickly realize that this work can be complicated to do alone, hence the importance of discussing it with your team. Each brain has its own way of working.

By the way, all this does not only apply to IT, clean code, and PHP. That is why it was mentioned in the previous chapter that clean code is not just a set of rules: it is a way of being—it is a mindset.

Being consistent – get results quicker

Being perfectly consistent in what you do will force you to understand what you are doing. Then, everything will become a habit. If you have these good habits to the point that they have become natural to you, results will come faster in two specific cases, as outlined here:

- As we have seen from the beginning, you will be able to understand each other much faster within your team—developers will have the same habits. More rarely, but it can happen: you will sometimes have to discuss code or show things to non-technical people in your project. Although these people—such as a product owner, for example—may have some basic technical knowledge, it is best to assume that you will need to go back to the most basic basics. You will have a much easier time explaining a complex and technical subject to someone non-technical if you have done things simply, cleanly, and without hesitation in your work.

- The second case is during automated checks. Automated checks are tasks that have been set up and discussed with the team and that are executed every time you want to propose changes. These checks can take place in several places. It can be in your software to write code (IDEs such as NetBeans, PhpStorm, or **Visual Studio Code (VS Code)**), through **continuous integration (CI)** tools (GitHub Actions, GitLab CI/CD), and so on.

Automated auditing tasks can include absolutely any task you want. We will go into more detail on how to perform these tasks as soon as *Chapter 7, Code Quality Tools,* of this book (in the *Code quality tools* section), but here are the most common ones:

- Check code style and indentation

- Run test suites (unit, functional…)

- Run a static analysis of your code to make sure that the variables you use are well defined, that they use the right types, and so on

- Set up alerts to notify you of the success or failure of tasks through a chosen channel (email, instant messaging such as Slack, and so on)

- Deploy to a test environment

- Install dependencies and vendors

- Copy files to the server and do some remote operations

- Whatever you want!

You get it. These tools allow you to perform the tasks you want and need. In reality, they are just orchestrators that will play the commands you define. It is as simple as that. After that, if your commands are complex, it is another story. But you realize how infinite these automated checks can be.

About source code analysis tools

By having good code habits, you will definitely speed up the process. Perhaps the most telling and concrete example of this is style-code checking. If you do not know how to write code in your team and your team has put automatic checks on it for everything you want to change, you may spend hours figuring out that a space was missing in one place, a line break in another, and so on. Do not worry—most tools offer options to correct these errors automatically.

However, this is not the case with static analysis tools, such as the ones we're going to set up in the section dedicated to static analysis tools in *Chapter 7, Code Quality Tools*. In fact, a static analysis will review your code and make sure that the most common errors are not made. We are not talking about checking the number of spaces in your tabs, but really parsing the PHP to try to understand it and make sure everything is in order. These tools can be a bit too strict at times, and it can take a lot of time to understand them properly. Also, these tools are not perfect, and the static analysis tool may not be able to understand what you want to do. Although this is a separate issue, you will save yourself a lot of trouble if you deal with it at the source: develop good coding habits. Be thorough and do not leave anything to chance. PHP is a very lax language that allows you to do just about anything with variables—for example, allowing you to typecast without flinching. As we know, the results of this kind of operation can be random, even very surprising, and may seem completely illogical. Anyway, PHP is what it is. Although static analysis tools can spot these risky cases most of the time, you may spend hours correcting these small things that can number in the dozens very quickly.

Also (and this may sound silly), the cleaner you code by implementing clean code practices and what follows, the fewer bugs you implement. By being sure of yourself, you save yourself a lot of scares. Moreover, you allow the next developers who will pass over your code not to be fooled too and make their passage easier. If you are lucky enough (or if you have applied the principles of clean code and what goes around it!), you will have tests that ensure the proper functioning of the application. Maybe you are not familiar with tests, so let's go through this together.

About testing and its multiple forms

Tests are mostly lines of code written by the developers. These tests make sure that for some input data, a specific output is returned. These **inputs and outputs (I/Os)** can be of diverse types and sizes. It can be an integer as well as a generated HTML page or even an image. To make things easier and to simplify the concept, automated tests are generally grouped into three main families, as follows:

- Unit tests, which are the tests with the finest granularity. They generally assess the return of the methods in the code while ignoring everything that surrounds them. What matters is the result returned by the function, and that's all.

- Functional tests have a medium granularity. They will assess full-fledged functionalities, where several parties and methods can be involved. The most obvious example is the testing of an **application programming interface** (**API**): we check that if we call a specific **Uniform Resource Locator** (**URL**) with specific parameters in the request, the API returns the expected result.

- **End-to-end** (**E2E**) tests are the most complex to maintain. These tests will mostly simulate a web browser, controlled automatically. A classic example is the test of a login form. A robot will automatically fill in the fields, click on the button, make sure that we are redirected and that a success message is present on the HTML page, and so on.

All these tests exist for a specific reason: non-regression.

Non-regression testing is without a doubt the thing that will save your position as a developer in a company. Alright—maybe that is a bit of an exaggeration. However, we cannot count the number of applications that have been saved thanks to it. When your test coverage is good enough (the proportion of lines of code and features covered by one or more tests), you can modify almost anything and be sure to never break anything if the tests are always *green*. Indeed, you can break features to rewrite them differently and test new ways of doing things. As long as the tests are green, you can be sure that the application behaves correctly, as it did before your modifications. Of course, several things come into play.

First, the tests have to test something. This may sound strange, but in fact, you end up with a lot of tests that do not actually assess anything. The most typical example is the unit tests on the setters and getters of a class. When you write tests for that, you are assessing that a variable assignment has been done and that a method call has been done. You are testing... PHP! And PHP already has its own tests. Writing relevant tests is a book in itself and is an art that can be mastered and refined over the years. And for that, there is nothing like practice—again and again.

The second thing to consider when you modify an application is simply that the tests you have set up are no longer up to date and must be modified. Sometimes, it can be difficult to understand whether a test fails voluntarily (that is, it becomes incompatible with your changes) or involuntarily (because the application does not behave the same way as before when it should). It depends entirely on your case. Remember one thing: if your application is tested correctly, you can change any line of code and deploy your application with your eyes closed, at any time, and without hesitation. Pretty interesting, isn't it?

Now that you see the benefit of tests in your application, we can discuss a practice that is quite common in the practice of clean code. You may have already heard of it: TDD.

TDD stands for **test-driven development**. It is a methodology that consists of writing tests before writing the rest of the code. At first, it is very confusing. It is complicated to understand how it happens, or even how it is possible. It is about thinking backward and questioning our thinking habits, yet the principle is quite trivial. First, we think about the tests—that is, the data we are going to send (to our methods in unit tests, to an API endpoint in the case of functional tests, and so on), as well as the output we want (a precise object or value in unit tests; a precise **JavaScript Object Notation** (**JSON**) return, for example, in the case of an API functional test). Obviously, and because you have not written the rest of the code yet, all tests fail. This is intentional. The goal is now to make these tests go green one by one.

If you try this practice, you will realize that a kind of magic happens without even realizing it. You will organize your code in a way that you would never have done at first. It will be cut up in a clear and precise way so that your tests can pass as quickly and easily as possible. In addition to the exceptional intellectual satisfaction that will result, you will end up with instantly more readable and cleaner code. Also (and contrary to widespread belief and intuition), developments will be much faster. There is indeed a period of adaptation for it to become quite natural, and you may have the impression of being awfully slow at first. However, thanks to this, your code is simpler and therefore faster to write, understand, adapt, and extend. You have to try it to experience this, as it may sound a bit miraculous. And it really is in some way.

Moreover, the code coverage becomes increasingly extensive as you develop. This has a direct impact on the maintenance of your application, as we said before: you will be much more confident in amending your code, as well as all the people who will have to read and modify it. The tests will be there to protect you. As a bonus, reading tests can be invaluable in getting into complex code. By reading the tests, you can understand from the I/Os given where the developer writing the tests was going. This is priceless in the vast majority of applications, especially legacy ones. Also, a lot of developers are looking at tests in the first place when reviewing your code. It is an amazing entry point when getting into someone's change to the code base. See your tests as the one and only way to prove that what you just did actually works. This is a reality: most developers sensitive to clean code and alike consider tests as the only valuable proof that your changes work.

It can be noted, for example, that for most (if not all) open source projects such as PHP, you must add tests when you make changes. Whether it's for adding a new feature or fixing a problem, testing will be mandatory, and your changes will never be approved without testing. These tests will be, once again, irrefutable proof of the behavior of your new feature or that you have indeed fixed the bug in question. All this is a lot of work, but with this in place, the code coverage becomes huge, and you contribute fully to the stability of the software.

Testing is invaluable, both for the quality of your code and for the speed with which you get results. Thanks to it, you will be consistent, and you will get results quicker.

Summary

We have just seen a lot of new knowledge together. If you understand it, you can be sure that you are already a better developer than you were in the previous chapter.

Knowing the SOLID principles is a real asset in the professional world and in industrial-quality projects. Even if each case is different and each project has its specificities, these principles have the advantage of being applicable almost everywhere, and at least of being very strongly inspired by them.

Keeping in mind the KISS, DRY, and YAGNI principles will allow you to keep your feet on the ground and not spread yourself too thin during your next developments. They emphasize thinking about the present moment to help prepare for the future, rather than thinking about the future to try to adapt to the present moment. You should definitely remember this. We don't know the future constraints that will be imposed on us, whether technical or functional, so it makes more sense to think about how to make it easier to deal with those constraints rather than guess at them. Because let's face it: we have very little chance of hitting the bull's eye.

If you have the opportunity and the possibility, implementing the "Scouts' principle" in a TDD strategy can be more than beneficial and will always be an excellent idea. If you have never practiced TDD, and despite the explanations given in this chapter, it is quite normal to be completely dubious about its usefulness and—especially—its effectiveness. This is normal, and we have all been there. However, the results are there, and the various case studies that have been conducted on this subject demonstrate this drastically. It might be time to test this way of doing things, which is very well seen and appreciated by the seniors of clean code!

In spite of all this, we must keep in mind that clean code is also about adapting to its environment. It is not a question of rewriting the application entirely and changing all the habits of the development team under the pretext that someone outside the project has decided to do so. You must be aware of your context and deal with your environment. You must be able to adapt to the need and to what is around you. This is what will make you a good "clean-coder." Remember to communicate as much as possible with your team when a change of habit is perceived, and be able to justify all your choices. If possible, it will always be good to propose several solutions, as well as the advantages and disadvantages of each.

After all this theory, we can move on to a slightly more practical part. What are the ways to write clean code? What is the purpose of code? Although we have seen some advanced principles, we should not forget the basics, and we should also question what we already know.

3
Code, Don't Do Stunts

The advanced principles of clean code will actually help you to become an easily understood developer who is able to code more cleanly. They teach you to be consistent in your choices, to think about other developers and your team, and to make communication the main tool of our work. Even before the source code.

It's a fact: although the source code has a preponderant place in the developer's job, we should not make it our main reason to be. It is a reality: the developer's job is not to write code. It's about finding a solution to a given problem while adapting to constraints that would get in the way. This is the basis of our job that we must absolutely keep in mind. And although the principles we have just seen in the previous chapter, such as SOLID, seem to be strongly linked to the code, we must try to have a more *"meta"* perspective on all this, thinking outside the box, and taking a step back. The principles mentioned are, objectively, tools that will allow us to solve the problems posed in an efficient and straightforward way.

We can then ask ourselves the following questions: What is the real purpose of source code? What is its purpose, and can we allow ourselves to do anything with the most basic things in the language?

These are the topics we'll be covering in this chapter:

- Understanding code
- Be understood, not clever
- A note on maintainability

Understanding code

Let's first ask ourselves about the importance of code. What really is its importance for us developers in our daily lives? For that, let's go back in time.

A bit of history

Computer programming is, in fact, a transistor through which an electric current passes or does not pass. So, we end up with a binary system, with a value of 0 if the electric current does not pass through the transistor, and 1 if the electric current does pass. If you multiply this number of transistors by several billion, you end up with today's processors. It works very well, and our world has been governed by this system for decades. However, there is a clear limitation: it is not humanly possible to understand and create applications with only 0s and 1s. So, we had to find a new way of writing these programs so that they became humanly possible and manageable.

We then move to the first human-readable source code: the **assembly** language (often abbreviated **ASM**). Popularized at the end of the 1940s, this language finally made it possible to read files with a language that is more or less similar to our natural language, although assembly is a very low-level language (meaning that it is remarkably close to the language of the machine—namely, binary). One thing led to another, and higher-level languages appeared—the C language being the best known, seeing its first official version in 1972. The principle is simple: to be able to write computer programs with a language that is more and more natural for human beings. A tool then automatically translates this higher-level language into assembly and binary language that the machine can interpret.

The C language is an excellent example of the usefulness and main purpose of a programming language. Indeed, this language, created by Dennis Ritchie and Brian Kernighan, was originally used to develop the Unix operating system. The point is that it was easier to create a programming language such as C to write the Unix operating system than to write the operating system with the tools of the time—namely, assembler (even if some parts of the Unix operating system are written in assembly, the vast majority of the source code is written in the C language). We'll look at the actual purpose of programming languages and code next.

The purpose of code

And this is where it all makes sense. The programming languages are there to help us to transcribe our ideas as easily as possible, and they are interpretable by the machine. But programming languages are not limited to transcribing our ideas to machines—their goal is also that other people can understand our ideas by reading our code, and without needing us. Programming languages are a subtle mix of the potential to be understood by a human while giving us the leeway to communicate with the machine and the possibilities to exploit its full potential. This is what defines the level of a programming language when we talk about high-level and low-level languages: the position of the cursor is between "ease of use and understanding" and "performance and possibilities offered by the language". There are necessary trade-offs to be made. These are things to have in mind when choosing the most appropriate programming language when kicking off a new project. A perfect example is all the tools bundled in the **PHP: Hypertext Preprocessor** (**PHP**) language to deal with **Hypertext Transfer Protocol** (**HTTP**) requests and responses, which makes it an excellent choice to create a web application. Most of the things you will need are already present, out of the box, and without the need to install anything to deal with the most basic and some pretty advanced web application features.

PHP is written in C; it is a higher-level language than the C language. It is, therefore, easier to understand and more permissive but offers, in comparison to C, less extended performance and fewer possibilities. If you ever need to write some assembly code in PHP to communicate with some custom hardware, for instance, the statement is simple: you cannot. You will have to write a PHP extension, which will be written in the C language (which then allows you to write source code parts in the assembly language). This is a super-advanced case, of course, but you get the point.

Let's go further by comparing PHP and **Hypertext Markup Language** (**HTML**). Although HTML is not a programming language but a description language, it still has similarities with PHP: both languages are used to express human ideas that can be interpreted by the machine. The point is simple: for someone who doesn't really know technical and programming languages, you will certainly be able to explain the content of an HTML file, what it represents, its semantics, and its purpose. In PHP, it's a different matter. Indeed, between file splitting, class splitting, **object-oriented programming** (**OOP**), and all these other concepts, it will certainly take you much more time to make your non-technical interlocutor understand the purpose of all this. However, although HTML does not allow OOP, it does not allow conditional branching, writing to a file, managing requests sent to the server, and so on. So, we end up with a language that is much more understandable to a human being because it is very close to our natural language but with much fewer possibilities.

Despite these differences, we must keep in mind the following thing—the languages' main objective is exactly the same: to be understood by the greatest number of people and by the computer. Writing code means being understandable. It is to expose ideas. And just as in everyday life when you expose your ideas, the simpler and more straightforward you are, the more people will be able to understand you.

Be understood, not clever

It happens very often that in front of technical challenges and especially in source code, we want to do things in a fine, pretty, even "sexy" way, as some would say. This is completely normal—since code is a major part of our lives as developers, we sometimes want to show the extent of our abilities. Although this can be justified at times, it is often an awfully bad idea to want to show the complete extent of these talents. Obviously, our ego takes a hit—we must hold back sometimes. You just learned new ways of doing things, new ways of coding, and new principles of which you are ardently convinced. You have spent a weekend learning this new way of organizing your code and your project, you experience it as a revelation, and you are sure of it: you must show this new discovery to your colleagues and your team; it will revolutionize the project and bring only good things. Moreover, you will be given the credit for this new thing, and you will be the referent of it. However, this is not the right approach at all.

Don't be mistaken. Learning every day, on your own time or not, is an exceptional thing. If you have the opportunity to do so, you will come out better. On a side note: this isn't mandatory in any way! Nothing should force yourself to code in your spare time. It's totally OK to keep coding and programming for work.

The desire to share your discoveries and experience with your peers is normal—even more than that, it's healthy. Sharing knowledge to lift your loved ones up is the best thing ever, and explaining something is the best way to learn yourself. The mistake is to want to apply it immediately, from everywhere, all the time. Each way has its advantages and disadvantages. It is absolutely necessary to be aware of the disadvantages that it brings. In general, the most common ones are the application in the current project and the resistance to change from other people involved in the project. Just take the example of the famous SOLID principles: although their effectiveness is proven, they can be difficult to access for newcomers.

Resistance to change is normal and natural for all people. Our brain likes regularity—it likes cycles and does not like the unexpected. This is obviously reflected in the work environment but also in all other aspects of life: diet, exercise, and sleep. For code and our work habits, it is exactly the same. Again, consistency and regularity are key.

If you bring your new discoveries into the project, you're going to have to consider the training of the people already in place. They will not necessarily want to change their habits if the habits already in place are appropriate and already meet the need. New habits also mean training all the people who do not know these ways of doing things. This requires personal investment, even a substantial one in some cases. It is then a question of learning hours either in personal time or during working hours, and these hours will therefore be hours where productivity will be reduced, undeniably. Sometimes this is necessary, and sometimes it's even a promising idea. But then, you must be able to justify it to everyone, including the non-technical parties in the project, and this can clearly be a critical part, especially if you are pushed for time.

Also, some programming practices can work wonders, be proven, and make life much easier. However, they have a huge disadvantage: the onboarding time on the project. Just take as an example the practice of "if-less programming". This programming method says to never, absolutely never use `if` and conditional branching in your code. This calls for massive and pure use of **object-oriented programming** (**OOP**). On paper, it looks good, and the intellectual satisfaction of such a technique must be quite exceptional. Its efficiency once mastered is quite clear. Everything becomes smarter and the code becomes noticeably clear. In short, everything is in place for your next project to adopt if-less programming from the start.

However, the day when someone comes to your project to help you and try to understand what you are doing (or even to maintain and evolve the project with you), the observation will be overwhelming: if the person doesn't know this programming technique (being a technique far from a generality), the whole process of introducing the project will be painful. In addition to having to train the person in the functional constraints of the project, they will also have to be trained in a new programming technique that they are certainly not used to. This means understanding the project, understanding the stakes, changing one's habits, changing one's way of doing and working, and reshaping one's way of thinking. We understand very quickly the cost of such an operation. It can be justified, but you have to be very sure of yourself and know all the risks in advance.

We are talking about if-less programming here, but the same goes for other ways of doing things that are not a general rule. **Test-driven development** (**TDD**) is one of them! Integrating TDD into a project can be painful and complicated, as we have seen before. However, TDD mainly influences the order of doing things, more than learning a complete way of coding. It's up to you to see, depending on your context and your constraints, to what extent these risks are worth taking.

In any case, if you choose a new programming technique that can be described as exotic, you may have exemplary code, clean and efficient, and super-maintainable. The problem is that nobody will be able to understand it. Remember what was said in the previous section: code is used to express and convey ideas. It is used to be understood by machines and, especially, by human beings. It would be a shame to sacrifice the second point, which is why high-level programming languages were invented.

A note on maintainability

And this is where it gets complicated. Your code is ready—it works. You have followed a new programming method, and the preliminary developments of the project have been going smoothly for several months. And it's pretty obvious: there may have been no foundation on which to build your project; you were lucky enough to start from a blank sheet of paper. However, the question of maintainability will soon arise. Whichever programming technique you choose, whichever people are working on it, bugs will always appear. You may need new people to fix all this (and thus teach them your working methods). Are you sure that you have mastered your new methodologies enough to ensure the follow-up of an application over several years? It is quite possible, but you must be aware of this and know what to do if you get stuck on the maintenance of your application.

The purpose of this chapter is not to discourage initiative and the testing of new work methodologies. It is more a matter of being fully aware of the risks of opting for new working methodologies, especially in the long term. We will see later that we must be incredibly careful about the latest trends that can disappear overnight.

The same applies to programming syntaxes that may seem elegant at first glance but that are in reality a nightmare of maintainability. Among these practices, we can find, in a non-exhaustive way, those highlighted next.

Using binary operators and octal, hexadecimal, and binary notations

In general, the use of binary operators on integers to perform operations (left and right shifting, logical AND, logical OR, bit inversion, and so on) is more useless than anything else. Their rarity makes them a syntax that may seem elegant to perform certain operations. However, this is not the case, and mastering binary operations should not be a prerequisite for understanding PHP code.

The use of octal, hexadecimal, and binary notations can be justified at times. For example, octal notation can be used when you want to play with file permissions. Hexadecimal can be used if you want to use flags on your methods, as well as binary notation. But in general, apart from making your code complicated to read, there is not much to it.

Assigning a variable and using gotos

A variable can be assigned at the same time as testing its value. Here's an example:

```
if null === ($var = method()))
```

At best, you save one line of code. But it's been a long time since we cared about the size of application source files, which are optimized at runtime by the PHP interpreter anyway. There is no cost for assigning the variable before the test, and your code will become immediately more readable.

A goto instruction allows you to skip entire parts of the code, and even "go up" in the code. Although it can be useful in some extremely specific cases, it should definitely not be used in most cases. For years now, the use of goto statements has been frowned upon in most programming languages. Indeed, they bring great complexity to the understanding of the code flow. There is a name when too many uses of goto are to be deplored: spaghetti code.

Excessively using comments

Sometimes, we see the abusive use of comments on sometimes several hundred lines to explain the type of all the variables, their utility, all the exceptions raised by the function, the return value in detail, and so on. We can even find source files where there are more comments than code. All these details can be omitted most of the time by a clear naming of your methods and variables. Furthermore, the typing of variables, arguments, and function returns in recent versions of PHP also solves this problem. However, using comments to produce documentation can be totally justified and should be used when possible. Nobody will ever complain because there is "too much documentation". Feel free to write dozens of lines about what a class, interface, method, and so on is all about; more generally, descendant, technical and/or functional choices about it, and so on. When we speak about "abusive use of comments", we're talking about comments that are explaining what's happening in the code.

Using ternary comparisons

Here's an example of a ternary comparison:

```
$var === null ? 'is null' : 'is not null'
```

Although ternary comparisons can make code concise and include a condition in—for example—the passing of an argument in a function, they should not be abused, especially nested ternary comparisons, which immediately become unreadable and headache-inducing at the first level of nesting. You can see proof of this in the following example:

```
$var === null ? 'is null' : is_int($var) ? 'is int' : 'is not
null'
```

This is not a very clear or readable line of code, which makes ternary conditions complicated to read when the condition is not a basic and simple one, as in the first example.

Using abbreviations

Here's maybe the most common practice to be discouraged: the use of abbreviations all over the place. Again, at the end of the last millennium, we might have had reasons to use abbreviations: space and storage were much more limited than today, and code editors were not as smart with all the autocomplete features we have today. Therefore, naming a variable `$userPasswordRequest` instead of `$usr` will make everyone's life easier: both yours and the developers who will come back to your code and won't have to ask you what your abbreviations mean. Again, with the autocomplete tools we have today, it doesn't make sense to name our variables this way.

Bringing micro-optimizations to your code

Micro-optimizations are very minor changes made to the code that may impair its readability, justified by the optimization of the code. Thus, it would be faster to execute. The problem is that this is often not useful, first because you don't need to optimize an instruction to the nanosecond (because of the power of processors nowadays), but also because a lot of optimizations are done by language interpreters and compilers. So, you sacrifice part of the readability of your code for something that is not useful and that will be done automatically. Moreover, this often provokes sterile debates where nobody is more right than anyone else. In these debates of micro-optimization specific to PHP, we find in particular the position of the operators of incrementation (++) and decrementation (--) in position before or after the variables, the use of the backslash in front of the methods of the **Standard PHP Library** (**SPL**), or the static declaration or not of anonymous functions. Again, the answer is: think about consistency with the rest of the code and be pragmatic. You certainly don't need the 10 nanoseconds that will be (perhaps, in some circumstances) saved by your optimization that will have started a heated debate within the development team.

Recoding the methods of the SPL

We have a lot of use for a language that has a very extensive standard library. The standard library is a set of classes and methods provided with each installation of PHP. Unfortunately, we quickly realize that it is quite unknown and offers more possibilities than you might think. As a result, we often find ourselves with SPL methods that are recoded in the project because the developer in question did not know about the existence of the standard method. This is extremely unfortunate and, in some cases, a real problem for the following reasons:

- SPL methods are available everywhere. There is no need to worry about whether they are available on this or that installation or setup.

- These methods are tested by the developers of the PHP interpreter, which is not necessarily the case for your methods.

- If one of these methods can be optimized or secured, it can be done thanks to the thousands of contributors and researchers of the language.

- These methods are thought and conceptualized to be as efficient as possible by people whose job is to create the most efficient algorithms possible.

- SPL methods can be written in the C language directly. This means that their performance will be unmatched no matter what you do in PHP. It would be a shame not to take advantage of this considerable benefit, especially on methods used intensively in applications where execution time can be critical. Furthermore, because they are written in C, the C compiler can offer very low-level optimizations on these methods, directly with assembly code. You won't be able to do this by writing the method in PHP.

Feel free to have a look at the official PHP documentation; some methods such as `natsort()` might surprise you and save you hours of development!

The list could go on and on, but the point is that while you may enjoy using these things, you will be the only one who will feel any satisfaction. A junior developer might be completely lost at the sight of these practices, while a senior developer won't understand the value of using these practices when clearer and simpler ones are available. Your code must be understood by as many people as possible. Show the extent of your skills, knowledge, and proficiency by producing some simple, trivial, and readable code for a problem that seemed overly complex at first.

Summary

Do you understand what is meant by "building on what we know"? We realize with a little hindsight that this is again a lot of common sense and altruism, thinking of the next developers who will pass over our code. Here, there was no question of scouts, SOLID, **Keep It Simple, Stupid (KISS)**, or other principles. It's about rethinking our very methods of writing code.

We must remember that the basics can (and should) be questioned and not considered as set in stone. Self-confidence is a wonderful thing, and if you are able to combine this with continuous questioning of your habits in a perspective of continuous improvement, you are on the right track to becoming an excellent developer, being able to write clean code naturally, and bringing your collaborators with you in this practice.

Taking the initiative is a remarkable thing; knowing the risks and evaluating them in your context is the key to striving for perfection. This way, you are able to know if it is really worth it, but also can justify your choices to the people running the project. Again, we come back to the ability to justify all our choices and actions when it comes to clean code. Clean code is not just about avoiding the use of binary operators or the use of hexadecimal notations. It means thinking about the environment, the constraints, and the surroundings of our project. Clean code is not only about code. Luckily, this is exactly what we will see in the next chapter.

It is about More Than Just Code

Wouldn't describing **PHP: Hypertext Preprocessor** (**PHP**) as a programming language be a bit reductive when you think about it? We must face the facts: PHP is not a simple programming language. It's a complete ecosystem, with a gigantic community, thousands of contributors, and new features being proposed and released regularly. But not only that: millions of libraries and **application programming interfaces** (**API**) are written and launched thanks to PHP. Even many command-line tools are entirely developed thanks to the PHP language. PHP is a whole world on its own. Let's start by looking at the reasons why PHP is not just a language for writing a website.

These are the topics we will cover in this chapter:

- PHP as an ecosystem
- Choosing the right libraries
- A word about semantic versioning
- Stability versus trends

PHP as an ecosystem

This can be seen in several things that we can list together, as follows:

- PHP is, still in the early 2020s, the most used server-side language for web application development. When you know the predominant (not to say overwhelming) place of web applications in our everyday use, this is a genuinely nice award!

- The language continues to evolve very strongly, especially in recent years. It went through a slump during the development of PHP 6 (which was never released) before experiencing a real explosion of its popularity starting with version 7. Version 7 defined the foundation of the future of PHP with highly demanded features such as strong typing, as well as incredible performance and speed improvement. Benchmarks comparing PHP 5 and 7 were just crazy when they first came out. Developments continue strongly, with new features being proposed very regularly.

- PHP has an exceptional dependency manager named **Composer**. Simple, open source, and devilishly efficient, it is often recognized by its users as the best dependency manager on the market, all programming languages included. Although this may be a subjective opinion, we can't take away its reliability.

- Speaking of dependencies, you only have to visit the Packagist site (the repository where Composer takes dependencies) to realize the exceptional community that PHP has at its disposal to make available so many libraries, each one more incredible than the other, with the vast majority being free of charge and with no usage restrictions. If you have a need, there is an external library available that will certainly solve your problem.

- PHP extensions are a real goldmine for extending the language. The extensions, unlike the libraries that you install with Composer, are written in C and plug directly into the source code of the PHP interpreter. This gives us the possibility to extend the language with impressive performance. This also means that the default installation of PHP can be extremely minimal and needs almost nothing to work. We can then install extensions unitarily according to our needs. This is especially useful if our application must run on a server with limited resources.

- Multiple conferences around the world show a strong commitment to PHP. Equally incredible and renowned tools such as the Symfony, Drupal, and Laravel frameworks show a real desire to push the language as far as it can be taken. These frameworks themselves organize international conferences and are used by multinational companies (Airbnb, Spotify, TheFork, and so on). For the record, Symfony is still in 2022 one of the open source projects with the largest number of contributors to the framework and documentation, among all existing open source projects.

- The development of the PHP core is still in full swing today, and more than ever. Proposals for new features and **requests for comments** (**RFCs**) (the first step in proposing a change to the language) are emerging at a rapid pace and are implemented just as fast. Many contributors are involved, and a renewal of the main contributors is also observed. Older and more important contributors, such as the well-known Nikita Popov, are leaving the ship while new contributors are coming to the project. The community is in perpetual effervescence.

PHP is a language that has a proven reputation. Its robustness and efficiency have made it the language of choice for some of the world's largest sites. Where alternatives have been implemented in the past or currently, such as Python (which is used by around 1.2% of all websites at the time of writing) for some Google sites, PHP is in the majority. Obviously, many attractive technologies are emerging and taking market share from PHP, such as Node.js or C# and the .NET Framework. PHP still has a good future ahead of it. Knowing how to write a website in PHP ensures that you'll know how to read the source code of the overwhelming majority of existing sites in the world.

For all these reasons, PHP is an ecosystem. Pushing the reflection a little further... if PHP is not just a programming language and if PHP is not just code, why should we limit clean code to code?

The clean code could also contain, by extension, the right choice of external dependencies and libraries to install on your project. Let's see why you should choose your dependencies wisely and how to choose them well to limit the risks.

Choosing the right libraries

Choosing the right external library to install can be a real challenge. It's a challenge we've all faced or will all face one day. The reason is simple: there is no point in reinventing the wheel. The reason we want to install an external library is usually the same. We have a specific problem that we want to solve as cleanly as possible. Here, two situations arise:

- We know how to solve the problem, but we don't want to have to rewrite everything when tools already exist to solve our problem simply

- We have no idea how to solve the problem because we lack theoretical or practical knowledge

It is then interesting to call upon an external library whose role is to bring us a very specific solution to our case. The advantages are multiple, as outlined here:

- The person(s) who develop(s) the external library may have thought for several days or weeks about the best way to provide a solution. It may even be their day job. Whatever time they have spent on it, it is often more time than we will allow to think calmly and cleanly about the solution.

- The maintenance of the dependency, if it is still actively in development, is managed by someone else than you. This means that you will regularly receive bug fixes and new features thanks to these volunteers who offer you their most precious asset: their time.

- If the external library you are using is open source, then you are exposed to additional benefits, as follows:

  - The source code is visible to everyone. This means that anyone, such as other developers or even security researchers, can analyze the source code in order to strengthen it and fix security flaws (or at least notify the author). Don't get me wrong: open source is not bad for security. In fact, it is quite the opposite. Security by obfuscation (understand this as hiding things such as source code to ensure "security") is the worst thing that can happen. Security must be achieved by other means. There is no lack of evidence: the most used encryption and ciphering algorithms in the world are known, and their functioning is perfectly explained in 1,001 places on the internet. This does not compromise their security.

  - If the open source project is abandoned, then initiatives (called "forks") can follow. Forks are, to put it simply, people who have copied the source code of a project and developed it on their own, independently of the developments of the original project. This can ensure, in theory, infinite longevity of a project.

  - If the main maintainer of the project no longer has time to take care of the project but developers wish to do so, the source code being open to all, they can do so.

- If you are curious, you can dig into the source code and understand how the problem was solved by the library!

We can clearly see that the choice of open source dependencies is quite inevitable. If you want the insurance to not end up with an unusable tool unavailable overnight, open source is made for you because you can store a copy of the source code as long as you want without fear. This is the first excellent way to choose an external library.

A second factor to consider is the frequency of updates to the project. Obviously, if a project is open source but has not been updated for several months or even several years, beware: it may be an abandonment. In this case, it means that the project may not support the next versions of PHP, for example, or that the bugs and security flaws will not be fixed anymore. There are two effortless ways to know if a project is still maintained or not, as set out here:

- First, you can check the date of the last version of the library. Be careful again, as some projects have (very) slow-release processes, and it is advisable to combine this technique with the second one.

- Here's the second technique. Look at when the last modifications of the source code were made. This can be done very easily, especially if the source code is hosted on a site such as GitHub. By browsing through the files, you can see when the last modifications of a folder or a file were made. This can be an excellent indication of the development dynamics of the project.

A third factor to consider is the documentation of the library. You'll probably want to make sure that the project has minimal and sufficient documentation to set up the basics. If no documentation is provided, you can be sure that using the external library will be a systematic pain. Indeed, all code maintenance will become a battle to remember how the project works, without documentation to help you or to share knowledge. Moreover, this can also be very much related to the community around the technology you want to use. If very few people use the project you want to integrate into yours and the community is quite inactive, or even non-existent, nobody will be able to help you in an optimal way. This can be an effective way to decide whether to use this or that dependency in your code.

A fourth factor is the number of dependencies that the library itself depends on. Generally speaking, we prefer a library that has very few dependencies. Fewer dependencies mean fewer packages to update and fewer third parties, so there is less chance of problems in one of those parties.

Finally, many projects have **continuous integration** (**CI**) badges on their main page. These badges allow you to know at a glance the test coverage (as a reminder: the proportion of code covered by tests), the number of tests, the latest version, and so on. Obviously, it is better to choose a project with as many tests as possible and with a high test coverage to limit problems during updates.

A word about semantic versioning

Speaking of updates, let's talk about versioning and—especially—semantic versioning. If the external library you want to use follows the rules of semantic versioning, this could have an incredibly positive and reassuring impact on your developments and updates. Let's take a look at what this means exactly.

What is semantic versioning?

Versioning is simply putting a number on a version of the source code. We are all familiar with versions such as 1.0, 1.5.0, 2.0.0, and so on. The semantic versioning adds a semantic—that is to say, precise meaning to each of these numbers. Let's take version 2.3.15 as an example. Here is how semantic versioning breaks down this version number:

- The "2" indicates a major version. A major version can introduce new features, bug fixes, and—most importantly—changes that break backward compatibility. This last point is the most important. Indeed, from one major version to another, method signatures or even complete class names can change, and some may also disappear. So, you have to be extremely careful when you move to a major version higher than the current one, and you have to test that everything still works. Often, the release changelogs provide the changes you need to make to be compliant with the new major release.

- The "3" indicates a minor version. As with major releases, minor releases can bring new features as well as bug fixes. The main difference is that minor releases cannot make changes that break backward compatibility. This means that you can upgrade a dependency to the next minor version to take advantage of all the new features and bug fixes without worrying that your code will break when you upgrade. However, it will never be superfluous to run all your tests afterward at the time of the minor update. You never know. Often, minor releases trigger code deprecation messages. These messages tell you which methods you should not use anymore because they will certainly be removed in the next major release. By taking into account the deprecation messages as you develop, you save yourself a lot of work when it comes to updating the dependency to the next major release.

- The "15" indicates the patch number. A patch contains only bug fixes and security fixes. It does not contain new features. You should consider *always* installing the new patches of your dependencies in your project.

We can see the advantages of semantic versioning: serenity, logic, and consistency. There are obviously other variations such as Alpha, Beta, Release Candidate, and Golden Master. But these are rarer.

How to deal with semantic versioning

Semantic versioning also includes a particular notation that allows your dependency manager to know how to install new versions and when to update your dependencies. Let's take as an example this extract of the file that Composer uses to install dependencies (and this is the same principle for many dependency managers out there):

```
{
    "require": {
        "php": ">=7.3",
        "symfony/dotenv": "3.4.*",
        "symfony/event-dispatcher-contracts": "~1.1",
        "symfony/http-client": "^4.2.2"
    }
}
```

This snippet describes four dependencies: a minimal version for PHP, as well as three external libraries. It doesn't really matter what these libraries are. We can note here four separate ways to define the versions we want to accept in our dependencies. Let's see what they are.

The first way to write the version we observe is by using the >= operator. This one is one of the easiest to understand: we want to accept all versions greater than or equal to the one specified. Here, our application accepts all versions of PHP higher than version 7.3, as well as version 7.3 itself. Of course, dependency managers accept other such operators: =, <, >, and <=. You can also combine these operators to get very precise version constraints—for example, by writing ">=1.2.0 <2.0.0".

The second operator is quite well known because it is used in many other contexts. It is the wildcard, denoted *. This symbol simply represents the fact that you can replace it with whatever you want. In the preceding example, we accept all the patch versions of the 3.4 version of the dependency. This allows it to benefit only from bug fixes, without updating the minor version. This wildcard can be placed anywhere in the version number. For example, the notation 3.* will benefit from all minor versions of the major version 3.

The following notation is the use of the tilde operator, ~. This operator means that you will only benefit from the patches of the given version. In the example, we will then benefit from all the patch versions of version 1.1 of the dependency (that is, 1.1.0, 1.1.1, 1.1.2, and so on). This is remarkably similar to the wildcard operator, except that the wildcard operator cannot be placed anywhere in the version number and only concerns patches. Also, it is worth noting that Composer interprets the tilde a little differently: it also allows minor versions, not just patches. If you are using Composer and you want to benefit only from the patch versions without the minor versions, you will have to use the wildcard operator.

Finally, the last operator we will see is the caret operator, denoted ^. In the preceding example, the caret operator allows all patch versions as well as minor versions of major version 4 (that is, 4.2.2, 4.2.3, 4.4.0, and so on). If you want to define a minimum version of a dependency while accepting new patches and minor versions but refusing major versions (which may bring breaking changes) automatically during the update of external libraries, this is a particularly good choice. That's why it's one of the most popular operators.

The possibilities are endless, and once you have mastered this notation, you can be confident about updating the dependencies of your project. As far as good practices are concerned, it is always a clever idea to accept all new patches and minor versions of a dependency. You should never lock a dependency to an extremely specific version without any conditions or possibility to update. Indeed, if an external library scrupulously respects semantic versioning, you will have no conflict with your existing code. Breaking changes are reserved for major versions. Therefore, you should not automatically accept major versions when updating your dependencies: chances are that you will have to adapt your code to make it work properly.

Stability versus trends

Let's finish this chapter with a few words about the most recent versions, but also about trendy external technologies and libraries.

First, let's talk about the latest versions of external libraries. Of course, we might be tempted to use the latest ones, the ones that were just released a few hours ago. It is worth remembering that bugs may appear, and a new patch version may be released in the near future if this is the case. Or not. And in this case, the bug could persist for a while. So, it's particularly important to write tests. Imagine the comfort: you update all your dependencies, you run your test suite, and if all the lights are green (and your application is properly tested), you can be fairly sure that everything is fine.

That said, if any tests turn red because you've updated an external library, you'll have to investigate to find out where this is coming from. In any case, you shouldn't think that you are safe from any problem if your dependencies are well fixed and constrained or you only accept patches and/or minor versions. Patches could also bring bugs—you never know.

As far as Alpha versions are concerned, let's be clear: these versions are not made for production applications. The different libraries are clear on this point: the code can change from one day to the next, bringing breaking changes without warning. In short, you must be incredibly careful. That said, if you want to evaluate these versions to see for yourself, the developers of the libraries will be delighted to receive your feedback. The Beta versions are supposed to be more stable and not bring any more breaking changes. You should still be incredibly careful when using them.

As a general rule, only use the final, stable versions in production. Reserve the Alpha and Beta versions for development and test environments if you want to be ready on the day of the stable release for production deployment. New features are always exciting things, but they are never worth sacrificing the stability of your application. Your users don't care about the new features of the external libraries you use: only stability matters—the fact that it just works.

Now, let's talk about trendy technologies (an external PHP library, a new tool, or even a new programming language). You hear everyone around you talking about a particular technology. This technology is spreading like wildfire, you hear about it everywhere on the internet, huge companies are getting into it, and tech conferences are all about it. You must be wary of this kind of thing. Even if the promises of these technologies can be exciting and revolutionary, think first about what is important: your users.

Will this technology make a real difference to your end users? Is it really worth training on it and taking weeks or even months to figure out how it works? You need to be sure that it will have a real positive impact on your project. You also must keep in mind that innovative technology will have a small community. The impacts are immediate, as outlined here:

- You will have to train all the people who arrive on your project

- The documentation may not be complete, which may make it difficult to understand

- You may find yourself alone in front of your screen without finding a solution to your problem: you are one of the first to use this technology, and therefore one of the first to face the obstacles encountered with it

Finally, you must make sure that the project is robust so that you don't end up with a recent technology abandoned without warning. This happens more often than you think, and much (if not all) of your work will have been for naught. So, beware of the latest unproven technologies, and be sure of the robustness and seriousness of the project. Wait until you've had some time to think about it. Again, your users will surely be able to do without this technology (which they will not be aware of) until it is mature.

Summary

Limiting PHP to the programming language is reductive. We have just seen it—it is a real ecosystem with a rich and active community, and extremely far from burying its favorite language. The developments around PHP are countless, and the language itself has evolved in the most beautiful way in recent years. The contributions of functionalities gave a real second wind to this one, allowing it to claim—still today—first place among the most used programming languages on the server side for a web application.

All this would be nothing without the explosion in the number of external libraries available for the language. You have a problem; there is a solution. We are fortunate that most external libraries are open source. Thousands of developers make available, voluntarily and free of charge, the fruit of hours, weeks, or years of work.

Making a choice from among these libraries can be difficult and challenging. It is important, even mandatory, to do real research work beforehand to be sure to make the right choice. We are not immune to obstacles and incidents, but this chapter has provided you with tools and ready-to-use solutions to limit the risks. Above all, don't rush into the most fashionable technologies. If you want to attract users and have them continue to use your application more than another, the key words are "robustness" and "stability"!

We have talked a lot about other people's work, but we should not forget our own achievements. How can you manage to develop good habits to find your way in your code as you manage to find your way effortlessly in the source code of your favorite external libraries when you need to understand its internal workings? We come back to what we said in the first chapters: by having the same habits, we understand each other more easily. This obviously applies to the organization of a project, in the naming of files, the structure of folders, and so on. And this is exactly what we will see in practice in the next chapter.

5

Optimizing Your Time and Separating Responsibilities

It's time for a little practice after all that theory! We've already seen a lot together: advanced principles on clean code, how to choose the right external libraries for your applications, and how to take advantage of the latest patches of these libraries while not risking making your project explode in mid-air. But we should not forget that there is the word "code" in "clean code" (obviously). In this chapter, we will therefore concentrate a little more on the source code of your application and see the following points:

- Naming conventions and organization of files and folders

- Why is it important to separate responsibilities to respect the "S" of the SOLID principles? What does it bring to you?

- We'll discover an elegant way to manage responsibility separation with an event system

- And we'll finish with some polymorphism—namely, abstract classes and interfaces: why, how, and when to use them?

Naming and organizational conventions

We must add a disclaimer before anything else. The naming conventions and organizational ideas given in this chapter are not an absolute truth. As we have seen before, the most important thing is to respect the conventions already in place in your project and to be consistent with your team. If you feel it is necessary, it is possible to adapt these rules to your needs. Again, the important thing is to use common sense and logic and to be as clear as possible.

Let's first talk about the naming of source files. Obviously, the naming conventions differ from one technology to another (depending on whether you use a certain framework or another, the good practices may change, for example). Nevertheless, we can note some conventions that can be found almost everywhere.

Class files and interface files

PHP: Hypertext Preprocessor (**PHP**) source files defining a class, an abstract class, or an interface should have the same name as the class or interface in question. For example, the `Foo` class should be defined in a `Foo.php` file. More than a convention, this naming technique has a real technical interest. Indeed, the autoloading mechanisms of PHP will assume that your file defines a class with the same name. Autoloading allows PHP to automatically discover the classes defined in your application, and in particular thanks to namespaces (we'll come back to this in a few moments, as these are directly linked to the organization of files in your project). If you name your files and the classes they define differently, autoloading is likely to fail and throw an error. The most common naming style across the various languages and the global developer community for naming classes is **PascalCase**. This style simply adds a capital letter to the beginning of each word. So, a class named *My super service class* would be named `MySuperServiceClass` if we used PascalCase. There are other naming styles—we will see some of them and in which cases they apply.

Executables

PHP files, being executable command-line scripts, have a strong tendency to be named in lowercase. The **PHPUnit** test framework is a good example. In practice, it is used as a system command in a terminal. In reality, it is simply a PHP script. We use the hyphen as a word separator. This naming style is called **kebab-case**. This is the naming style that is mostly used by command-line programs. The well-known `apt-get`, `docker-compose`, and `git cherry-pick` are perfect ambassadors. Nothing prevents you from naming your executable file in another way, and everything will work fine. However, by naming your command-line applications written in PHP this way, you provide a uniform **command-line interface** (**CLI**) experience with the vast majority of commands. This is exactly what we want when we develop a command-line application in PHP: for it to blend in with more *traditional* system commands.

Web assets and resources

There is another case where the kebab-case style is used, which is mainly for public web resources, particularly JavaScript, **Cascading Style Sheets** (**CSS**), and **Hypertext Markup Language** (**HTML**) files. Indeed, kebab-case is a naming style easily readable and understandable by everyone, whether you are in the technical field or not. Moreover, search engines are better able to understand the semantics of your resource by using this notation. It is therefore essential if you want to do **search engine optimization** (**SEO**). Here's an example: it's much more common—even natural—to see a **Uniform Resource Locator** (**URL**) such as `/contact-us`, compared to something such as `/contactus` or `/ContactUs`. This can be particularly important if you must work on frontend files.

Naming classes, interfaces, and methods

As we have seen in the previous chapters, abbreviations should be banned from your code. **Integrated development environments** (**IDEs**) are nowadays powerful enough to autocomplete the names of the classes and methods you use. Therefore, it is not useful to shorten repeatedly the names of your classes, methods, and variables as this only brings about confusion. This is also why your abstract classes should start with the prefix `Abstract` (for example, `AbstractMailer`) and your interfaces should end with the suffix `Interface` (for example, `MailerInterface`). This makes the name very slightly longer, but there is no confusion in their use. Their purpose is clear, defined, and visible immediately. Don't be afraid to give your classes long names if it is necessary for their understanding. `AbstractWebDeveloperConsoleStreamWrapperExtension` may seem extremely long for a class name, but it is immediately clear what it is for in a project context, without having to ask too many questions. Again, with the autocomplete feature of your IDE, you will be able to use it within seconds by typing the first few letters. The same goes for your attribute and method names. Be explicit.

Talking about attribute and method naming, we tend to prefer the **camelCase** naming style. The principle is the same as for PascalCase (that is, put a capital letter at the beginning of each word), except that the first letter of the name is in lowercase (for example, `myGreatMethod`). Some languages use PascalCase for naming methods such as C#. Let's be honest—there is no real justification or argument for this, and both naming styles are actually equal. For once, it is really a convention—a language-specific habit.

Naming folders

The naming conventions for folders are similar to those for files. PascalCase is mostly used for folders. Other naming styles can also be used for publicly exposed folders, such as web resources. You should not hesitate to create a large tree structure and give it meaning. Therefore, folders named simply `Manager`, `Service`, or `Wrapper` are to be avoided. These terms are too generic and do not allow an easy understanding of what they define. We prefer more explicit variants, such as `Mail` and its subfolders `Provider`, `Logger`, and so on. These subfolders can have more generic names, being contained in a folder whose name defines a context: a domain. A clever way to find your way around is to use folders to separate your source code into different domains. Your application will be better sliced and your architecture clear. What concerns the same theme will be in the same place. You will be much more efficient as this habit becomes natural. Many open source projects and libraries use this way of slicing their sources. So, you will be able to browse not only your own code but also the code of others. It is sometimes extremely complicated and time-consuming to find the right name for an element, whether it is a class, a file, a variable, or anything else. This step is, however, particularly important and should not be neglected. Be careful not to say to yourself: "I put a name that is not necessarily very clear, but I will change it later." Chances are that you will forget it, and technical debt will be brought in without you even realizing it.

Separating responsibilities

Let's see what the separation of responsibilities in the code consists of, to make it cleaner, understandable, maintainable, and extensible. This is the first point of the SOLID principles. In the second chapter, this is how we defined the principle of single responsibility: "*It means that a class in your code must respond to only one task.*"

As a reminder, SOLID is a set of known clean-code rules that, when applied together, will make your code much clearer and more accurate. Rather than trying to follow the five principles described by each of the SOLID caps to the letter, it is more important to have a global idea of all this in mind when you code.

The first step to respect this is, in fact... naming, as we just saw! Indeed, by naming a class properly, clearly, and, above all, precisely, you are already making sure that it doesn't become a mess where you put a little bit of everything that you can think of. And it is for this very reason that it is necessary not to name your methods with too generic terms such as `Manager` and `Service`. This leads to a big problem: if we end up with a class named `EmailManager`, we will obviously all have as a first thought to add all our next methods that deal with managing emails. And that's how the chaos begins. That's why we will prefer to create classes such as `EmailFactory`, `AbstractEmailSender`, and so on in order to absolutely avoid having classes with hundreds of different methods.

We start to understand better this principle of single responsibility. Let's repeat: the goal is not to create classes with a single method in them. It doesn't make sense. You must split it up intelligently. There is no general rule for splitting classes. The right way to split a class will come naturally with experience and it will come by itself. If it helps you, you can see folders as domains, and the files as subdomains. Using the next examples, we have a domain (or folder) that would be named `Email`, and subdomains dedicated to specific tasks: creating an email, defining a base class to send an email with a specific email provider, and so on. We can go even further in this separation of responsibilities. Indeed, tools exist to help us solve this problem in an effortless way. We are going to discover (or rediscover!) event dispatching.

Event dispatching

Event dispatching is usually implemented thanks to the Observer and Mediator design patterns, as is the case in Symfony's `EventDispatcher` component. This information is just for general knowledge. Indeed, design patterns can seem obscure, even scary at first. Moreover, explaining them deserves a book of its own. So, we're going to vulgarize all this without talking about design patterns. Moreover, we are not going to implement an event dispatcher: it is about understanding how it can help us.

Very simply, the principle of event dispatching is to notify all parties interested in the change of state of a particular entity when it happens. The parties will notify a central mediator by saying: "I'm interested in knowing when this particular event occurs; notify me when it does because I have things to do if it happens." The mediator will then retain this information. When said event occurs, the mediator will then go through the list of parties interested in it and say: "The event just happened; do what you have to do." Taking this a step further, it is even possible that the interested parties will declare a priority to go before everyone else if necessary. Be careful—we are talking about synchronous events, which is to say that the parties interested in the event will execute one by one, following the others, and not in parallel. Asynchronous event management is a whole other story.

Event dispatching is as simple as that. But then, how will it help us reinforce our principle of single responsibility? Let's look at a concrete example: a user deletes their account from your application. You then need to do two tasks, as follows:

- Delete the account from the database
- Send a last email to the user to say a sad goodbye

So, we will naturally create a service such as `UserRemover`, which will perform these two tasks in a row. It works very well. `UserRemover` is an explicit name that defines a very precise task. No problem so far. Then, one day, your application gains popularity. You want to send an email to the administrators to notify them that the user has left. Our `UserRemover` class ends up deleting data and sending two emails, both with very specific content and recipients.

Later on, you want to give the user the possibility to delete their account to respect the **General Data Protection Regulation** (**GDPR**) because you are based in Europe. Only, you want to have accurate statistics and keep anonymized data of the user to respect their choice to disappear from your application while having the possibility to exploit their anonymous data to improve your product. You then have an `UserRemover` service that actually removes nothing, sends emails, and maybe many other tasks. We have a real problem: the class doesn't do what it is supposed to do, and it is likely that it is now thousands of lines long and has dozens of methods.

Things would have been hugely different if you had used event dispatching from the beginning. Here is an example of a resolution. When a user wants to leave your application, you will dispatch an event named `UserRemovalRequestEvent`. Then, as your application grows, you will create parties interested in this event: event listeners. We will have one per task, as follows:

- A listener to remove data from the database
- A listener for the goodbye mail to the user
- A listener for the mail to the administrators

What about anonymization? Nothing simpler: we will create a listener for this task too, and we will "unplug" the listener for deleting data. So, we have one class per task. Each class has its own unique responsibility (send an email to the administrators, anonymize data, and so on). If in the future you need to add a task, you will create a new class (or listener) with a specific task, without ever touching the other classes. The code is then clean and extremely extensible. The classes remain well-named and concise. If a task is obsolete, you can simply remove it from the list of parties interested in the event in question. The principle of single responsibility is respected.

If you want to use a ready-to-use event dispatcher, we recommend using the `symfony/event-dispatcher` package. This is exactly the component that is used in the framework for its operation. It is very robust and efficient and has been proven for several years.

Demystifying polymorphism – interfaces and abstract classes

As far as the separation of responsibilities is concerned, event dispatching is a concept that is already advanced. You can consider that your level in the world of clean code has increased considerably if you know this mechanism, understand it, and have the chance to use it. All this obviously requires a bit of setup. Either you implement this system by yourself or you use an external library. In the second case, there is obviously a whole learning phase to be included. Either way, this is obviously not the only way to improve the separation of your responsibilities. There is a way that is native to PHP to improve this separation, sometimes not used enough, sometimes misunderstood, and often underestimated. We are talking here about polymorphism, or vulgarly: abstract classes and interfaces.

First, why the word *polymorphism*? *Poly* comes from the Greek meaning *many*, and *morphism* means *form/shape*. Abstract classes and interfaces are just one way to implement polymorphism in code and **object-oriented programming** (**OOP**). To simplify things, let's take only the case of interfaces.

Interfaces

Interfaces define a common form/shape for classes that implement them later. They determine the methods that each implementation should define for its case. An implementation must necessarily define all the methods of its interface(s). This is often the reason why we hear the following statement: "an interface is a contract". We can see it this way: if you implement an interface, you commit to implementing the methods it defines. You have no other alternative.

This is where the full power of polymorphism takes place. In your code, you can tell PHP that a so-and-so argument of a method will necessarily be an instance of an object implementing a precise interface. You can manipulate the different methods of this interface and call them. You don't have to worry about how it will be implemented when it is used: we are not interested in that. As an example, is worth a thousand words, let's take our mail system.

You have `MailerInterface` that defines only one method: a method to send a mail. We could name it `sendEmail`. As before, when a user is deleted in your application, the event listener for sending a goodbye mail is called. In this case, what you are interested in is that the mail is simply sent, not the internal workings of the sending. By the way, couldn't these internal workings be different depending on certain conditions? You don't have to look far for an example: your main email provider could be down, but you absolutely need to send your message. You then must use another email provider, with a different **application programming interface** (**API**), different options, and so on. Without polymorphism, things could get extremely complicated very quickly.

The solution is to create two implementations of `MailerInterface`, each defining the `sendEmail` method, depending on the email provider used. But the result is the same: the mail is sent. When a user deletes their account, you perform a check to make sure that your main email provider is up and instantiate its implementation. If it is down, you instantiate the backup email provider implementation. On the other hand, in the email sending event listener, you just keep calling the `sendEmail` method defined in `MailerInterface`, without worrying about the rest. The code is clean, clear; responsibilities are separated; you save time. And on top of that, it has become resilient to failure.

You could do this with 10, 15, or 20 email providers if you wanted to. The advantage is that if the API of one of the providers changes or you find a bug in your implementation, you only have to touch the one implementation that is problematic. All the others don't move, as is the case with the calls that are made to the interface. You considerably reduce the risk of bugs, and your code is much more testable: you can write tests specifically for each implementation. This is much more robust than generalized, endless tests that try to evaluate every possible case! The time saved is exceptional and priceless.

Abstract classes

And how do abstract classes fit in all this? We can see these as an intermediate layer between an interface and its implementations. Although it is obviously not mandatory that an abstract class implements an interface, it is often a clever idea to create an interface on top of an abstract class. Indeed, the latter are more permissive than interfaces: you can partially define the methods that are declared, declare attributes, and decide on the visibility of the methods and attributes of the class (where interfaces only allow `public` visibility and not `private` or `protected`). With an interface, you have a clean contract, without any information that would be here "just in case" and with only the bare essentials if you respect the "I" of the SOLID principles—the interfaces' segregation. As a reminder and simply put, this principle indicates that an interface should not contain methods that are declared "just in case" and that each one should help to meet the principle of single responsibility.

Abstract classes allow us to define common behaviors to the classes that extend them and that wish to take advantage of the power of polymorphism. In particular, this avoids code redundancy, sources of bugs, and endless copy-pasting. Indeed, in our previous example, it is highly likely that the different implementations of `MailerInterface` have common behaviors, such as the creation of the HTTP client that communicates with the API of the email provider or the creation of a `Message` common object used in the internal workings of the implementations.

In this case, we would declare `AbstractMailer` implementing `MailerInterface` and define common behaviors of the different implementations. Then, the different implementations would extend `AbstractMailer` in order to enjoy the common behavior you just defined.

Be careful—this does not mean that it is necessary to create interfaces and abstract classes everywhere, all the time, and in all cases. We should not neglect the impact that this has on the complexity of the code compared to a single class. Also, just because you haven't created an interface for a case doesn't mean that it is immutable and set in stone. Very often, we find ourselves refactoring our code and creating interfaces and abstract classes and adapting existing classes to implement and extend them. As we have seen, we need to keep the code simple at first (respecting the *YAGNI* and *KISS* principles). We cannot predict the future, business constraints evolve.

If at the time of creating a class, there is nothing to suggest that different implementations will be needed, this is not a concern. This is a job that will be done later. On the other hand, if during development you find yourself copying code from one side to the other and perceive a strong redundancy, it will be an excellent reflex to think about polymorphism.

Summary

We have just covered the most advanced part of the theoretical section of this book. We are now armed with the knowledge to cut our code cleanly while keeping it maintainable and extensible for future developers. It will also be ready for the future by being strongly open to extension and closed to modification (as described in one of the *SOLID* principles).

We have reviewed many of the cases that you may encounter regarding the naming of files, classes, and methods when developing a PHP application. In addition, we have seen that folders must have specific names and can be used to divide your application into different domains.

The separation of responsibilities was also a big topic. It is particularly important to understand why this separation is useful, even vital, in a project. It is the real key to a well-architected project that is easy to navigate. Event dispatching is an excellent way to achieve this, as we have seen. Event dispatching is one of the cornerstones of some critical web projects such as the Symfony framework, to name but one. This one relies heavily on this mechanism, making it a tool known for its robustness, its efficiency, and—especially—its flexibility. This is also due to the polymorphism and the different interfaces declared within it. You can redeclare pretty much any part of the framework thanks to this and adapt everything to your most advanced needs.

It is not always easy to understand when to create an interface or an abstract class. This comes with practice and experience. Soon, it will seem natural. When in doubt, talk to one of your peers!

We will finish the theoretical part of this book with a lighter part in the next chapter, dealing with the new features of PHP. These make us more rigorous and better developers, especially in the last few years.

## 6

PHP is Evolving – Deprecations and Revolutions

We did well. The PHP community did well; we were lucky. Indeed, PHP has been evolving very strongly for a few years now. But this strong evolution hasn't always been there. This was mainly due to problems during the development of PHP 6, which was why this version was never released. This explains why so many projects were (and still are) stuck at PHP 5.

PHP 7 has wiped the slate clean and brought a real revival to the language. Moreover, it is a real breath of fresh air that has boosted the language toward new horizons.

PHP went from being an almost dead language to a language catching up and projecting itself in the future. In this last chapter, dedicated to the clean-code theory, we will focus on the following points:

- How PHP is different from its past versions
- How these changes will help you become a more rigorous and better developer, and not only in PHP
- What the major new features of PHP are in its latest versions

Old versus new PHP

PHP has likely helped you become a much more rigorous developer over the years. If during its first decades of existence, PHP allowed you to write code the way you wanted to and without restricting you from doing so, with the (very) few advantages that this brings, in hindsight, it was mostly the opportunity to have as many ways to write code as there are developers (which rarely lead to exceptional results) that made it popular. As we now know, that can be a source of endless and infernal bugs to debug. Fortunately, the evolution of the language in the last few years has fixed a lot of these bugs, to the benefit of our applications.

Strict typing

First, let's look at one of the most important things you should be using in the newest versions of PHP from version 7.4 – the strict typing of properties.

There was a time you were allowed to pass any data to any variable and cast variables as much as you wish, without a real and native way to prevent this – converting an array variable to a string, a string to an integer, and so on. That can be pretty confusing and potentially the source of so many problems. What would happen if you multiplied a string by an integer, for example? Well, the result is totally unexpected. This isn't well managed like it is in Python, where these types of operations are allowed but well controlled. If your PHP code relies on the ability of weak typing, there may be a problem with your code architecture, and you must absolutely review the parts that need this weak typing.

PHP now allows you to strictly type variables in some situations. As of PHP 8.1, you can't type a variable if it is not a class attribute or a method argument. That said, you should type all the attributes of your classes and method arguments – less confusion, more rigor. You may have to rethink some parts of your code, but you get the benefit of cleaner, more understandable code. There are no surprises at runtime because of an unexpected cast. If you really need to pass any type of data to your methods, you can still rely on the `mixed` keyword, which tells PHP that this variable can be any type of data, or that the method can return any type of data. Of course, you must avoid this if you can and only use it in very precise cases (such as interface method definitions, where the interface's implementations can return several types of data).

Error reporting

PHP 8 now displays deprecations and more strict errors by default. The error reporting level was lower in earlier versions of PHP. With this change, you'll be able to see way more easily where you need to pay attention to deprecation, for example. You'll be thankful for this change when you have to operate a PHP version upgrade, for example. If you take care of fixing deprecations as soon as they appear, upgrading to another version of PHP will be an easy task. Moreover, there is almost always a message resulting from the deprecation, telling you how to fix it precisely. Taking care of these errors and especially deprecations at the earliest opportunity is definitely a clever move and makes you adopt a clean-code mindset.

Attributes

Comments in source code were invented to give a better understanding of tricky code parts. This means that if we remove all comments in source code, it should work perfectly as well, as compilers and interpreters shouldn't consider comments. That's the main and *very first reason* source code comments were invented.

Then, annotations were created. Logic and mechanisms were introduced *right in the comment sections*. Don't get me wrong – annotations are very practical. You get every bit of information and metadata about an element where you need it and when you need it. But it is somehow an aberration when you think about it. And remember what we said in previous chapters: if you write code cleanly, then there is a great chance that you never need to write a single line of comment, or at the most just a few (writing tricky code parts cannot always be avoided, even for the best clean coders).

Attributes have been part of the PHP language since version 8.0. Simply put, attributes have the same role as annotations: add metadata to different elements, including classes, properties, and methods. The difference is that they use another syntax, which is not the type used for comments. More than having better readability, comments will go back to their first use: being *informative*. In an instance, you can immediately differentiate metadata (described with attributes) and comments to better understand the code part you're working on.

Let's see what attributes look like:

```php
<?php

namespace App\Controller;

class ExampleController
{
    public function home(#[CurrentUser] User $user)
    {
        // ...
    }
}
```

It is clear that the currently logged-in user will probably be injected in the $user variable. The code looks sharp, and we have all the information we need to understand what's happening at a glance. We also have a perfectly blank space if we ever need to add information in the comments section above the home() method. You can now see clearly how attributes help you to be more rigorous – remove all comment blocks and think twice before adding a new one.

There are obviously still a lot of things happening in the most recent versions of PHP, starting from version 8.0. Here is a non-exhaustive list:

- Union types
- Match syntax
- Named arguments

- Enumerations

- JIT compiler

- Fibers

- Numeric separator

Let's see a few of them in the next section.

The version 8 revolution

As we have seen, PHP has experienced exceptional momentum in its evolution for the last few years. While we thought that version 7 was a real rebirth of the language, version 8 proved that it was only the beginning. Here are the main new features that will help you write clear and concise code, and that will help you to push even further the principles of clean code that we have seen throughout these chapters.

Match syntax

The match syntax is the condensed version of the classic `switch`/`case`. It should not be used everywhere because it can quickly become unreadable. However, if you choose the places where you use it sparingly, your code can become much clearer in an instant. Here is an example of the match syntax:

```
$foo = match($var) {
    <value 1> => Bar::myMethod1(),
    <value 2> => Bar::myMethod2(),
};
```

It works the same way as switch. However, note the difference in the length of the code and how readability is increased. You can also immediately see the limitations of such a syntax: if there is more than one statement to execute per case, it is not at all adaptable, and you run the risk of ending up with something unreadable. It is then preferable to stick to the more classical `switch`/`case` block.

Named arguments

If you are used to using other languages regularly, you may be familiar with this evolution of PHP. Since version 8.0, it is possible to pass arguments to methods in the order you want. This is done by specifying the name of the argument just before its value.

```
$this->myMethodCall(needle: 'Bar', enabled: true);
```

We can quickly see two cases where this can be useful:

- First, we sometimes encounter methods with a lot of optional arguments that already have a default value defined. Sometimes, we want to change the value of an argument that may be the 5th or 11th in a list. This is followed by a lengthy process of rewriting the default values of all the parameters before it. This is clearly not ideal, but it is a problem that can be solved by using named arguments. You simply specify the name of the argument you want to give and its value, and you're done.

- A second case is where you add clarity to a method call, even if it has no optional arguments. If your method has a lot of arguments, you might think about using named arguments to explicitly show what you are sending to the method. However, if this is the case, the real solution to make code more readable is to refactor it so that you don't need to pass so many arguments to a method. This can be done by splitting the method into several ones, or by creating **Value Objects** (**VOs**). These are objects containing only simple properties, without other methods or logic, to transfer data from one point to another in code. This avoids the worry of endless arguments, and it also adds a layer of validation with strong-typed properties and adds some context to the data you carry with it.

Read-only classes and properties

Here is a little riddle to introduce read-only classes and properties. In a PHP class, how can you ensure that an attribute will be assigned only once – that is, it can be assigned a value only once, and all future attempts to assign it will fail? Note that this also applies to assignments within the class itself. This means that a mutator throwing an exception in the case of a call cannot work because we cannot be 100% sure that the developer will go through the mutator within the class scope.

Actually, the answer is quite simple: it is impossible. If you are using PHP 8.0 or earlier, it just isn't possible. PHP 8.1 brings us a native solution to solve this problem: read-only properties. By declaring a class property with the `readonly` keyword, the variable can only be assigned once, regardless of the context. This can be especially useful when defining and using DTOs and VOs. By restricting access to the mutation of these properties, the developer using them will have to think carefully about the use of the object. If they want to make changes, they will have to create a new object. The original object cannot be modified, which can guarantee better stability and robustness of code. Here is how read-only properties are declared:

```php
<?php

namespace App\Model;

class MyValueObject
{
```

```
    protected readonly string $foo;

    public function __construct(string $foo)
    {
        $this->foo = $foo; // First assignment, all good
        // Any further assignment of $this->foo will result
          in a fatal error
    }
}
```

Since PHP 8.2, it has even been possible to declare an entire class as `readonly` by placing this modifier just before `class` keyword, so you don't have to use the keyword on every property the class includes. This also gives you a big advantage: a class declared as `readonly` will not be able to have new properties declared dynamically. Although this behavior is deprecated and should be avoided at all costs (dynamic declaration of class properties is a nightmare in terms of debugging, stability, and code complexity), it is still possible to do so in versions up to PHP 9.0, in which a fatal error will be triggered if this behavior occurs. If you are not using PHP 9.0, declaring your class as `readonly` will protect you from this behavior.

This is a big step forward for PHP developers. Indeed, all thanks to this feature, we will need to be, once again, increasingly rigorous on how we interact with objects and their properties.

Migrating resources to proper classes

Depending on how long you have been developing in PHP, you should be more or less familiar with what we call resources. A resource is a special type of variable, which represents a reference to an external resource. This may sound a bit vague, but it is actually quite trivial. A resource can be the following, for example:

- An open file
- A database connection
- A cURL call
- A connection to an LDAP directory (which is a way to manage user accounts in companies, generally)
- A font for a GD image manipulation extension

This has worked well for a few decades now, but we understand that the term *resource* is not really adapted anymore by being too generalist and especially quite old-fashioned now that classes and objects are predominant in modern code. Why aren't these resources simply objects like any other? Well, there is no particular reason (at least not at present) to justify this. And that's why PHP 8.0 has started a long but necessary migration to convert the old resources into full-fledged classes. It makes much more sense.

The use of resources was complex. They are complicated to debug and understand how they work and their internal state. They can only be called by special functions that handle resources. This is a big step forward for developers who will have more control and more tools to improve the rigor of development and robustness of their code.

The change is made as new versions of PHP are released. For example, PHP 8.0 embeds the migration of resources such as the following:

- GD, for the manipulation of images
- cURL
- OpenSSL
- XML
- Sockets

Conversely, PHP 8.1 embeds the migration of the following resources:

- GD fonts
- FTP
- IMAP
- finfo, for file management
- Pspell, for spell checking
- LDAP
- PostgreSQL

The following versions of PHP continue to do the same – to eventually remove the use of resources in the standard PHP library. Furthermore, the core PHP developers have decided since PHP 8.1 to create some of these classes under namespaces. We can see that the language, with this kind of action and development, is catching up after a long delay and making every effort to restore its reputation. PHP shows it clearly: the language is here to stay.

Protecting your sensitive arguments from leaking

If you have been developing with PHP for some time, you are surely aware of the many ways to display the content of a variable or a function argument. You can think of var_dump and print_r in particular. There are also other occasions when arguments and their values can be displayed: when the stack trace (or call stack) is displayed. This happens either by a manual call to a method such as debug_print_backtrace but also and often when an exception is thrown. In any case, if sensitive information is contained somewhere in the variables or call stack with the arguments of method calls, this can be problematic. You might think that this only happens in a development environment, but this is a mistake. It is highly likely that you write your exception messages in an error log somewhere on your server(s). This could result in sensitive information being displayed in your logs. This is obviously not recommended. Sensitive information should not be written in clear text anywhere. Also, and although it is a mistake, the security of application logs is often not as good as that of a database, for example. Furthermore, there is a great chance that many developers of the project (if not all of them) have access to such logs to debug the application. The threat does not always come from outside.

Fortunately, PHP 8.2 includes a new attribute to remedy this problem. It is indeed possible to indicate the #[SensitiveParameter] attribute before any function argument. This will tell PHP not to display the parameter value in var_dump, a stack trace, and so on. Placed cleverly, you'll be sure not to leak a sensitive value in an error message, for example. Indeed, it is not uncommon for sites to display the server error directly on the frontend. This should obviously be banned as soon as possible, but at least it helps to limit the damage. Let's see how this new attribute can be used:

```php
<?php

namespace App\Controller;

class SecurityController
{
    public function authenticate(string $username,
      #[SensitiveParameter] string $password)
    {
        // In case of any exception occurring or var_dump
           being called in here, the value of $password will
           be hidden in the different outputs
    }
}
```

In the internal workings of PHP, this attribute will replace the argument with an object of type `SensitiveParameterValue`, which will hide the real value of the argument. The argument will be displayed and present in the output, but its value will be hidden. Adding this attribute to your sensitive method arguments is a clever and welcome way to add rigor to your code and make it more resistant to attacks.

Summary

It cannot be repeated enough: PHP is evolving in the most beautiful way and is catching up with its competitors in the web world. The language listens to the community and the developers by offering them the tools they need to answer modern problems in the most viable way possible.

We have come a long way from a language that allowed everything and was very (too) lax for the challenges of today's web applications. Despite the explosion of frontend frameworks and technologies aimed at replacing server languages with languages intended for the frontend (such as Node.js with JavaScript), PHP has nothing to be ashamed of. Its impressive performance, its speed of evolution, and the reputation it has built over the years show that it still has a bright future ahead of it.

Although clean code is, as we have seen, a state of mind and, in a way, a philosophy, solutions native to the language are arriving in spades to help us apply them as well as possible. Even better, these features brought to PHP allow us to see new possibilities that we would not necessarily have thought of initially to make our code robust, maintainable, and viable in the long term. Just think about named arguments, read-only classes and properties, strict typing, or simply the last topic we covered in this chapter: protecting sensitive arguments from leaking into application logs and exception messages.

Having said that, it's time to get down to business. We'll start by looking at tools that will give you a very quick, numerical overview of the quality of your code. Having metrics for this kind of thing allows you to see what improvements you'll make to your code, or whether it's actually time to act because quality is diminishing as you develop. So, let's dive into the next chapter, which highlights code quality tools dedicated to the PHP language.

Part 2 – Maintaining Code Quality

The aim of the second part is to enable you to constantly improve your projects and eventually maintain a consistently high level of code quality. It will provide you guidance on using state-of-the-art tools and technologies, which will help to reduce the necessary efforts to achieve this goal. Finally, we will present several best practices that will help you to work together with other developers on a clean and maintainable code base.

This section comprises the following chapters:

- *Chapter 7, Code Quality Tools*
- *Chapter 8, Code Quality Metrics*
- *Chapter 9, Organizing PHP Quality Tools*
- *Chapter 10, Automated Testing*
- *Chapter 11, Continuous Integration*
- *Chapter 12, Working in a Team*
- *Chapter 13, Creating Effective Documentation*

7

Code Quality Tools

In the previous parts of this book, we learned the basics of clean code. Now, it is time to apply that knowledge to our everyday work. There are literally dozens of tools available for the PHP ecosystem that can help us detect flaws and potential bugs, apply the correct code styling, and generally inform us about quality issues.

To ensure a quick and easy start within the world of code quality tools, this section will introduce you to the most commonly used ones. For each, you will learn how to install, configure, and use it directly on your code. You will also learn about some useful extra features they provide.

We will look at the following groups of tools:

- Syntax checking and code styling

- Static code analysis

- IDE extensions

Technical requirements

For this chapter, you only need a bare minimum of tools to already be set up. The chances are high that you already have them installed if you have ever worked with PHP code before:

- A local installation of a recent PHP version (PHP 8.0 or higher is recommended).

- A code editor – often called an **Integrated Development Environment** (**IDE**).

- Composer, either installed as binary or globally. Please check `https://getcomposer.org/` if you are not familiar with Composer yet.

Please note that for the rest of this book, all examples are based on a Linux environment such as Ubuntu or macOS. If you are using Windows for development, you will most likely need to make some adjustments, as described here: `https://www.php.net/manual/en/install.windows.commandline.php`.

The code files for this chapter can be found here: `https://github.com/PacktPublishing/Clean-Code-in-PHP`

Syntax checking and code styling

The first group of tools we want to discuss helps us keep our code syntactically correct (i.e., it can be executed correctly by PHP) and formatted in a structured way. It seems to be obvious that the code needs to be written without errors, but it is always good to double-check, as some tools can actively change your code. Having a simple and fast way to ensure this will be essential when we automate the whole code quality process later in this book.

Having your code formatted following a common style guide reduces the effort required to read and understand your code, as well as the code of others. Especially when you are working in a team, an accepted style guide saves you hours of discussions about how to correctly format the code.

We will learn about the following tools:

- A PHP built-in linter
- The PHP CS Fixer tool

The PHP built-in linter

The first tool we want to look at is actually not a code quality tool of its own but rather an option built into the PHP binary itself: the Linter. It checks any code for syntax errors without executing it. This is particularly useful to ensure that the code works after refactoring sessions or when your code has been changed by an external tool.

Installation and usage

Since the Linter is already part of your PHP installation, we can immediately start using it by looking at an example. If you look closely, you will probably notice the error the author made in the following class example:

```php
<?php
class Example
{
    public function doSomething() bool
    {
        return true;
    }
}
```

Do not worry if you do not spot the error immediately – that is precisely what the Linter is there for! Simply pass the full name and path of the file to be checked to the PHP binary, using the -l option. By adding the -f option, PHP will also check for fatal errors, which is something we want. Both options can be combined into -lf.

Let us assume the preceding class can be found in the example.php file in the current folder – then, all we need to type is the following:

```
$ php -lf example.php
```

We will get the following output:

```
PHP Parse error: syntax error, unexpected identifier
  "bool", expecting ";" or "{" in example.php on line 5
Errors parsing example.php
```

You can tell the linter to check a complete directory as well:

```
$ php -lf src/*
```

> **Note**
> The built-in PHP linter stops on the first error – as in, it will not give you a full list of all the detected errors. So, you better make sure to run the command again after resolving the issue.

A recap of the PHP built-in Linter

The built-in PHP linter is a handy tool for quick code checks but cannot do much more than that. There are other more sophisticated linters such as https://github.com/overtrue/phplint. Not only will this one return a full list of errors but it can also run multiple processes in parallel, which will be noticeably faster on large code bases. However, other code quality tools already include a linter, such as the tool that we will check in the next section.

PHP CS Fixer: a code sniffer

Another essential tool is a code sniffer. It scans PHP code for coding standard violations and other bad practices. *PHP CS Fixer* (https://github.com/FriendsOfPHP/PHP-CS-Fixer) is a viable choice to start with, since, as the name already implies, it not only reports the findings but also fixes them right away.

> **Other code sniffers**
> *PHP CS Fixer* is not the only available code sniffer. Another well-known one is the *PHP_CodeSniffer* (https://github.com/squizlabs/PHP_CodeSniffer), which we can fully recommend using as well.

Installation and usage

Using Composer, the installation is straightforward:

```
$ composer require friendsofphp/php-cs-fixer --dev
```

> **Alternatives to Composer**
>
> There are multiple ways to install the tools we will introduce in this book. We will also check more options out later in this book.

The typical use case for code sniffers is to take care of the placement of brackets and the number of indentations, whether they're whitespaces or tabs. Let's check out the following PHP file with its ugly format:

```php
<?php
class Example
{
  public function doSomething(): bool { return true; }
}
```

If we run the code sniffer with its default settings, the command is nice and short:

```
$ vendor/bin/php-cs-fixer fix example.php
```

This will scan and fix the example.php file all in one go, leaving our code neat and shiny:

```php
<?php

class Example
{
    public function doSomething(): bool
    {
        return true;
    }
}
```

If you do not want to fix the file immediately, you can use the `--dry-run` option to just scan for issues. Add the `-v` option as well, to display the findings:

```
$ vendor/bin/php-cs-fixer fix example.php --dry-run -v
```

As with all code quality tools, you can also run it on all the files in a folder. The following command will scan the `src` folder recursively, so all subfolders are scanned as well:

```
$ vendor/bin/php-cs-fixer fix src
```

Rules and rulesets

So far, we used *PHP CS Fixer* with its default settings. Before we can change these defaults, let us have a closer look at how it knows what to check and fix.

A common pattern within code quality tools is the organization of rules within rulesets. A rule is a simple instruction that tells *PHP CS Fixer* how our code should be formatted regarding a certain aspect. For example, if we want to make use of strict types in PHP, every PHP file should contain the `declare(strict_types=1);` instruction.

There is a rule in *PHP CS Fixer* that can be used to force this:

```
$ vendor/bin/php-cs-fixer fix src
  --rules=declare_strict_types
```

This command will check each file in `src` and add `declare(strict_types=1);` after the opening PHP tag.

Since a coding standard such as PSR-12 (`https://www.php-fig.org/psr/psr-12/`) includes many instructions on how the code should be formatted, it would be cumbersome to add all these rules to the preceding command. That is why rulesets have been introduced, which are simply a combination of rules, and even other rulesets.

If we want to format code following PSR-12 explicitly, we can just run this:

```
$ vendor/bin/php-cs-fixer fix src --rules=@PSR12
```

As you can see, a ruleset is indicated by the @ symbol.

> **Rules and ruleset documentation**
>
> It is impossible to discuss every rule and ruleset for *PHP CS Fixer* within the scope of this book. If you are curious about what else it has to offer, please check out the official GitHub repository: `https://github.com/FriendsOfPHP/PHP-CS-Fixer/tree/master/doc`

Configuration

Executing commands manually is fine to start with, but at some point, we will not want to remember all the options every time. That is where configuration files come into play: most PHP code quality tools allow us to store the desired configuration in one or more files and in various formats, such as YAML, XML, or plain PHP.

For *PHP CS Fixer*, all the relevant settings can be controlled via the `.php-cs-fixer.dist.php` configuration file. Here, you will find an example:

```php
<?php
$finder = PhpCsFixer\Finder::create()
    ->in(__DIR__)
    ->exclude('templates');

$config = new PhpCsFixer\Config();
return $config->setRules([
    '@PSR12' => true,
    'declare_strict_types' => true,
    'array_syntax' => ['syntax' => 'short'],
])
->setFinder($finder);
```

Numerous things are happening here. Firstly, an instance of `PhpCsFixer\Finder` is created, which is configured to use the same directory to look for PHP files where this configuration file is located. As the `root` folder of the application is usually located here, we may want to exclude certain subdirectories (such as `templates` in this example) from being scanned.

Secondly, an instance of `PhpCsFixer\Config` is created. Here, we tell *PHP CS Fixer* which rules and rulesets to apply. We already discussed the `@PSR-12` ruleset, as well as the `declare_strict_types` rule. The `array_syntax` rule forces the usage of the short array syntax.

You may have noticed that the name of the configuration file, `.php-cs-fixer.dist.php`, contains the abbreviation `dist`. This stands for distribution and usually indicates that this file is the one the project gets distributed with. In other words, this is the file that gets added to the Git repository and is immediately available after checkout.

If you want to use your own configuration on your local system, you can create a copy of it and rename it `.php-cs-fixer.php`. If this file exists, *PHP CS Fixer* will use it instead of `dist-file`. It is good practice to let Git ignore this file. Otherwise, you might accidentally add your local settings to the repository.

Advanced usage

The ability of *PHP CS Fixer* does not stop at automatically fixing coding standard violations. It can also be used to apply small refactoring tasks. One great use case, for example, is the automated migration to a higher PHP version: *PHP CS Fixer* ships with migration rulesets, which can introduce some new language features to your code base.

For example, with PHP 8.0, it is possible to use the `class` keyword instead of the `get_class()` function. *PHP CS Fixer* can scan your code and replace certain lines – for example, see the following:

```
$class = get_class($someClass);
```

It can replace the preceding line with this:

```
$class = $someClass::class;
```

The migration rulesets are separated into non-risky and risky ones. Risky rulesets can potentially cause side effects, while non-risky ones usually do not cause any problems. A good example of a risky change is the `declare_strict_types` rule we discussed previously. Be sure to test your application thoroughly after applying them.

The capabilities of these migrations are limited – your code will not suddenly include all new PHP version features.

Code fixers cannot fix syntax errors for us. For example, the `Example` class that we checked with PHP's built-in linter in the previous section would still require the developer to manually fix it first.

Linting

PHP CS Fixer checks the files that you want to have sniffed for syntax errors as the very first step and will not apply any changes in case it finds syntax errors. This means that you do not have to run the PHP built-in linter as an additional step.

A recap of PHP CS Fixer

A code sniffer such as *PHP CS Fixer* should be part of every serious PHP project. The ability to fix rule violations automatically will save you many hours of work. If you chose not to apply any risky fixes, it will hardly cause any problems at all.

We have now learned how to ensure that our code is well-formatted and syntactically correct. While this is the foundation of any high-quality code, it does not help us to avoid bugs or maintainability issues. At this point, Static code analysis tools come into play.

Static Code Analysis

Static Code Analysis means that the only source of information is the code itself. Just by scanning the source code, these tools will find issues and problems that even the most senior developer in your team would miss during a code review.

These are the tools we would like to introduce you to in the next sections:

- phpcpd
- PHPMD
- PHPStan
- Psalm

phpcpd – the copy and paste detector

Copy and paste programming can be anything from simply annoying to a real threat to your projects. Bugs, security issues, and bad practices will get copied around and thus become harder to fix. Think of it as though it were a plague spreading through your code.

This form of programming is quite common, especially among less experienced developers, or in projects where the deadlines are very tight. Luckily, our clean code toolkit offers a remedy – the **PHP copy and paste detector (phpcpd)**.

Installation and usage

This tool can only be downloaded as *a self-containing PHP archive* (phar), so we will not use Composer to install it this time:

```
$ wget https://phar.phpunit.de/phpcpd.phar
```

> **Handling phar files**
>
> In *Chapter 9, Organizing PHP Quality Tools*, we will learn how to keep phar files organized. For now, it's enough to just download it.

Once downloaded, *phpcpd* can be used immediately without further configuration. It just requires the path of the target directory as a parameter. The following example shows how to scan the src directory for so-called "clones" (i.e., code that has been copied multiple times). Let's execute it with the default settings first:

```
$ php phpcpd.phar src
phpcpd 6.0.3 by Sebastian Bergmann.
```

```
No clones found.


Time: 00:00, Memory: 2.00 MB
```

If *phpcpd* does not detect any clones, it is worth checking the two options, `min-lines` and `min-tokens`, that control its "pickiness":

```
$ php phpcpd.phar --min-lines 4 --min-tokens 20 src
phpcpd 6.0.3 by Sebastian Bergmann.


Found 1 clones with 22 duplicated lines in 2 files:
- /src/example.php:12-23 (11 lines)
  /src/example.php:28-39
  /src/example3.php:7-18


32.35% duplicated lines out of 68 total lines of code.


Average size of duplication is 22 lines, largest clone has
   11 of lines


Time: 00:00.001, Memory: 2.00 MB
```

The `min-lines` option allows us to set the minimum number of lines a piece of code needs to have until it is considered a clone.

To understand the usage of `min-tokens`, we must clarify the meaning of a token in this context first: when you execute a script, PHP will internally use a so-called "tokenizer" to split the source code up into single tokens. A token is an independent component of your PHP program, such as a keyword, an operator, a constant, or a string. Think of them as words in human language. The `min-tokens` option therefore controls the number of instructions a piece of code contains before it is considered a clone.

You may want to play around with both parameters to find a good balance of "pickiness" for your code base. A certain amount of redundancy in your code is not automatically a problem and you also do not want to bother your fellow developers too much. Using the defaults to start with, therefore, is a good choice.

Further options

There are two more options you should be aware of:

- `--exclude <path>`: Excludes a path from the analysis. For example, unit tests often contain a lot of copy-and-paste code, so you want to exclude the `tests` folder. If you need to exclude multiple paths, the options can be given multiple times.

- `--fuzzy`: With this especially useful option, *phpcpd* will obfuscate the variable names when performing its check. This way, clones will be detected even if the variable names have been changed by a smart but lazy colleague.

Recap of phpcpd

Although *phpcpd* is easy to use, it is a significant help against the slow spread of copy and paste code in your projects. That is why we recommend adding it to your clean coder toolkit.

PHPMD: the PHP mess detector

A mess detector will scan code for potential issues, also known as "code smells" – parts of code that can introduce bugs, unexpected behavior, or are, in general, harder to maintain. As with the code style, there are certain rules that should be followed to avoid problems. The mess detector applies those rules to our code. The standard tool in the PHP ecosystem for this is *PHPMD*, which we will show you in this section.

Installation and usage

Before we take a closer look at what this tool has to offer for us, let us install it first using Composer:

```
$ composer require phpmd/phpmd --dev
```

After the installation is complete, we can run *PHPMD* already on the command line. It requires three arguments:

- The filename or path to scan (e.g., `src`). Multiple locations can be comma-separated.

- One of the following formats in which the report should be generated: `html`, `json`, `text`, or `xml`.

- One or more built-in rulesets or ruleset XML files (comma-separated).

For a quick start, let's scan the `src` folder, create the output as text, and use the built-in `cleancode` and `codesize` rulesets. We can do this by running the following command:

```
$ vendor/bin/phpmd src text cleancode,codesize
```

PHPMD writes all output to the standard output (stdout), which is on the command line. However, all output formats except text are not meant to be read there. If you want to get a first overview, you may want to use the html output, as it generates a nicely formatted and interactive report. To store the output in a file, we will redirect it to a file using the > operator as follows:

```
$ vendor/bin/phpmd src html cleancode,codesize > phpmd_report.
html
```

Simply open the HTML file on your browser and you will see a report similar to the one shown in *Figure 7.1*:

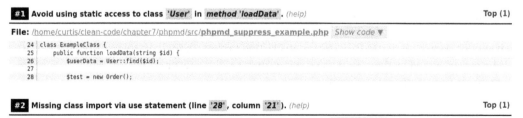

PHPMD Report

Generated at **2022-04-27 08:18** with [PHP Mess Detector](#) on **PHP 8.0.8** on **curtis-desktop**

2 problems found

Summary

By priority

Count	%	Priority
2	100.0 %	**Top (1)**

By rule set

Count	%	Rule set
2	100.0 %	**Clean Code Rules**

By name

Count	%	Rule name
1	50.0 %	**StaticAccess**
1	50.0 %	**MissingImport**

Details

#1 Avoid using static access to class *'User'* in *method 'loadData'* . *(help)*   Top (1)

File: /home/curtis/clean-code/chapter7/phpmd/src/**phpmd_suppress_example.php** *Show code* ▼

```
24  class ExampleClass {
25      public function loadData(string $id) {
26          $userData = User::find($id);
27
28          $test = new Order();
```

#2 Missing class import via use statement (line *'28'*, column *'21'*). *(help)*   Top (1)

File: /home/curtis/clean-code/chapter7/phpmd/src/**phpmd_suppress_example.php** *Show code* ▼

Figure 7.1: A PHPMD HTML report in a browser

The report is interactive, so make sure to click on buttons such as **Show details** or **Show code** to display all the information there is.

Rules and rulesets

In the preceding example, we used the built-in `cleancode` and `codesize` rulesets. Firstly, the rulesets are named according to the problem domain the rules check – as in, for the `cleancode` rule, you will only find rules that help to keep the code base clean. However, you can still end up with huge classes with many complex functions. To avoid this, adding the `codesize` ruleset is necessary.

The following table shows the available rulesets and their usage:

Ruleset	Short name	Description
Clean code rules	`cleancode`	Enforces clean code in general
Code size rules	`codesize`	Checks for long or complex code blocks
Controversial rules	`controversial`	Checks for best and bad practices where there are controversial opinions about them
Design rules	`design`	Helps find software design-related issues
Naming rules	`naming`	Avoids names that are too long or short
Unused code rules	`unused`	Detects unused code that can be deleted

Table 7.1: PHPMD rulesets

These built-in rules can simply be used by giving the aforementioned short names as arguments to the function call, as seen in the previous example.

If you are lucky enough to start a project on the green (i.e., from scratch), you can and should enforce as many rules from the beginning as you can. This will keep your code base clean right from the beginning. For existing projects, the effort is a bit greater, as we will see in the next section.

Using PHPMD in legacy projects

Often enough, you want to use *PHPMD* for an existing project, though. In this case, you will most likely be overwhelmed by the countless warnings that it will throw upon the first run. Do not give up – there are some options to help you!

Adjusting rulesets

If you plan to add *PHPMD* to an existing project, going all-in with the rulesets will surely lead to frustration because of how many issues are reported. You may want to concentrate on one or two rulesets at a time instead.

It is also highly likely that you will end up with rules that you find annoying or counter-productive at first – for example, the `ElseExpression` rule, which forbids the usage of `else` in an `if` expression. Leaving the discussion about the usefulness of this rule aside, the effort of rewriting countless statements that are working fine is not worth it. So, if you don't want to use that rule in your project, you need to create your own ruleset.

Rulesets are configured via XML files, which specify the rules that belong in them. Each rule is basically a PHP class that contains the rule logic. The following XML file defines a custom ruleset that just includes the `cleancode` and `codesize` rulesets:

```
<?xml version="1.0"?>
<ruleset name="Custom PHPMD rule set"
    xmlns=http://pmd.sf.net/ruleset/1.0.0
    xmlns:xsi=http://www.w3.org/2001/XMLSchema-instance
    xsi:schemaLocation=http://pmd.sf.net/ruleset/1.0.0  http://
pmd.sf.net/ruleset_xml_schema.xsd
xsi:noNamespaceSchemaLocation=
   "http://pmd.sf.net/ruleset_xml_schema.xsd">
    <description>
        Rule set which contains all codesize and cleancode
        rules
    </description>
    <rule ref="rulesets/codesize.xml" />
    <rule ref="rulesets/cleancode.xml" />
</ruleset>
```

XML seems to be a bit out of fashion nowadays, but it still serves its purpose well. You usually do not need to worry about all the attributes of the `<ruleset>` tag – just make sure that they are present. The `<description>` tag can contain any text that you deem to be a good description for the ruleset.

The `<rule>` tag is important for us. In the preceding example, we referenced both the `codesize` and `cleancode` rules.

> **Tip**
> At this point, it is a good idea to dig through the built-in rule sets in the GitHub repository
> `https://github.com/phpmd/phpmd/tree/master/src/main/resources/ rulesets`. Thanks to XML being a quite verbose file format, you will get familiar with it very quickly.

Imagine we want to remove the mentioned `ElseExpression` rule from our checks. To achieve this, you just need to add an `<exclude>` tag within the according `<rule>` tag as follows:

```
<rule ref="rulesets/cleancode.xml">
    <exclude name="ElseExpression" />
</rule>
```

This way, you can exclude as many rules from a ruleset as necessary. If you just want to pick certain rules from different rulesets, you can also go the other way round and reference the desired rules directly. If you want your custom ruleset to only include the `StaticAccess` and `UndefinedVariable` rules, your XML file should contain the following two tags:

```
<rule ref="rulesets/cleancode.xml/StaticAccess" />
<rule ref="rulesets/cleancode.xml/UndefinedVariable" />
```

One last important thing to know about the XML configuration files is how to change the individual properties of a rule. Again, a good way to figure out all the properties is to check out the actual ruleset file. Alternatively, you can check out the actual PHP classes of each rule at `https://github.com/phpmd/phpmd/tree/master/src/main/php/PHPMD/Rule`.

A typical example is to define exceptions for the `StaticAccess` rule. It is usually good practice to avoid static access, but often enough, you can't avoid it. Let us say your team agreed on allowing static access for the `DateTime` and `DateTimezone` objects – you can simply configure this as follows:

```
<rule ref="rulesets/cleancode.xml/StaticAccess">
    <properties>
        <property name="exceptions">
            <value>
                \DateTime,
                \DateTimezone
            </value>
        </property>
    </properties>
</rule>
```

To use this custom ruleset in the future, simply save the preceding XML in a file (usually called `phpmd.xml`) and pass it over to *PHPMD* upon the next run:

```
$ vendor/bin/phpmd src text phpmd.xml
```

> **Location of the configuration file**
>
> It is a common practice to place phpmd.xml with the rulesets you want to use in the root folder of your project and use it as the single source of configuration. If there are any modifications in the future, you only have to adjust one central file.

Suppressing warnings

Another useful tool for dealing with legacy code is the @SuppressWarnings DocBlock annotation. Let us assume one class in your project makes use of a static method call and that cannot be changed right now. By default, any static access will throw a warning. Since you do not want to use static access anywhere else in your code, but just in this class, removing the StaticAccess rule would be counterproductive.

In these cases, you can make use of the @SuppressWarnings annotation:

```
/**
 * @SuppressWarnings(PHPMD.StaticAccess)
 */
class ExampleClass {
    public function getUser(int $id): User {
        return User::find($id);
    }
}
```

You can use multiple annotations in one DocBlock if required. Finally, if you want to suppress any warnings on a class, just use the @SuppressWarnings(PHPMD) annotation.

Be aware that using the Suppress annotations should be your last resort. It is very tempting to just add it everywhere. However, it will silence the output, but it will not solve the problems.

Accepting violations

Instead of suppressing warnings at the file level or excluding rules from rulesets, you can also decide to acknowledge existing violations. For example, when you want to use *PHPMD* on a legacy project, you can decide to ignore all violations that are already in the code for now. However, if new violations are introduced by a new class, they will be reported.

Luckily, *PHPMD* makes this task quite easy by providing a so-called baseline file, which it will generate for you automatically by running the following:

```
$ vendor/bin/phpmd src text phpmd.xml --generate-baseline
```

In the preceding command, we expect that a `phpmd.xml` file already exists in the project `root` folder. Using the preceding command, *PHPMD* will now create a file called `phpmd.baseline.xml`.

Now, you may run the following:

```
$ vendor/bin/phpmd src text phpmd.xml
```

The next time, *PHPMD* will automatically detect the previously generated baseline file and use it to suppress all warnings accordingly. However, if a new rule violation is introduced in a new location, it will still be detected and reported as a violation.

A word of warning: as with the `@SuppressWarning` annotation, the baseline feature is not a tool that can be used once and safely ignored in the future. The problematic code blocks are still part of your project as technical debt with all the negative effects. That is why if you decide to go with the baseline feature, you should make sure you don't forget about addressing these hidden problems in the future.

We will discuss how to deal with these problems later in the book. For now, it is only important for you how to update the baseline file from time to time. Again, *PHPMD* makes this an easy task. Simply run the following:

```
$ vendor/bin/phpmd src text phpmd.xml --update-baseline
```

All violations that no longer exist in your code will be removed from the baseline file.

A recap on PHPMD

Unless you are starting a project on the green, the configuration of *PHPMD* will require a bit more time. Especially if you are working within a team, you will spend more time arguing about which rules to use and which to exclude. Once this is done, though, you have a powerful tool at your disposal that will help developers write high-quality, maintainable code.

PHPStan – a static analyzer for PHP

You might have noticed that *PHPMD*, which we looked at in the previous section, was not very PHP-specific but generally took care of the best coding practices. While this is, of course, very important, we want to use *PHPStan* to analyze our code with bad PHP practices in mind now.

As with every static analysis tool, *PHPStan* can only work with the information it can get out of the code. Therefore, it works better with modern object-oriented code. If, for example, the code makes strong use of strict typing, the analyzer has additional information to process, and will therefore return more results. But for older projects, it will be of immense help as well, as we will see in the following section.

Installation and usage

Installing *PHPStan* with Composer is just a one-liner again:

```
$ composer require phpstan/phpstan --dev
```

As with most code quality tools, *PHPStan* can be installed using PHAR. However, only when using Composer can you also install extensions. We will have a look at those a bit later in this section.

Let us use the following simplified example and store it inside the `src` folder:

```php
<?php

class Vat
{
    private float $vat = 0.19;

    public function getVat(): int
    {
        return $this->vat;
    }
}

class OrderPosition
{
    public function getGrossPrice(float $netPrice): float
    {
        $vatModel = new Vat();
        $vat = $vatModel->getVat();

        return $netPrice * (1 + $vat);
    }
}

$orderPosition = new OrderPosition();
echo $orderPosition->getGrossPrice(100);
```

To execute a scan, you need to specify the `analyse` keyword, together with the path to scan, which is `src` in our case:

```
$ vendor/bin/phpstan analyse src
```

Figure 7.2 shows the output produced by *PHPStan*:

```
$ vendor/bin/phpstan analyse src
Note: Using configuration file /home/curtis/clean-code/chapter7/phpstan/phpstan.neon.
 2/2 [                              ] 100%



 [OK] No errors


$
```

Figure 7.2: An example output of PHPStan

When we execute the PHP script, it will output `100`. Unfortunately, this is not correct because adding 19% of taxes to the net price should return 119, and not 100. So, there must be a bug somewhere. Let us see how *PHPStan* can help us here.

Rule levels

Unlike *PHPMD*, where you configure in detail which rules to apply, we will use different reporting levels here. These levels have been defined by the developers of *PHPStan*, starting from level 0 (just performing basic checks) to level 9 (being very strict on issues). To not overwhelm users with errors at first, *PHPStan* by default will use level 0, which only executes very few checks.

You can specify the level using the `level` (`-l` | `--level`) option. Let us try the next highest level:

```
$ vendor/bin/phpstan analyse --level 1 src
```

Using the level approach, you can effortlessly increase the quality of your code step by step, as we will demonstrate using the following, made-up example. Levels 1 and 2 will not return any errors either, though. As we eventually reach level 3, however, we will finally find a problem:

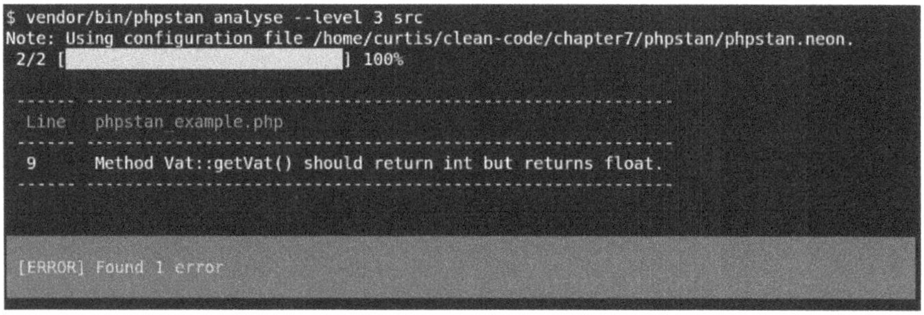

```
$ vendor/bin/phpstan analyse --level 3 src
Note: Using configuration file /home/curtis/clean-code/chapter7/phpstan/phpstan.neon.
 2/2 [                              ] 100%

 ------ --------------------------------------------------------------
  Line   phpstan example.php
 ------ --------------------------------------------------------------
  9      Method Vat::getVat() should return int but returns float.
 ------ --------------------------------------------------------------



 [ERROR] Found 1 error
```

Figure 7.3: PHPStan reports one error with level 3

Checking our code again, we can spot the problem quickly: the `getVat ()` method returns a float number (0.19) but using the `int` return type casts it to 0.

> **Strict typing**
>
> If we had used strict mode by adding the `declare(strict_types=1);` statement at the top of the example code, PHP would have thrown an error instead of silently casting the return value to `int`.

This demonstrates the beauty and power of Static Code Analysis: fixing this little bug will make our code work as expected and it takes us just a couple of seconds to do since we are still in our development environment. However, if this bug had reached the production environment, it would have taken us much longer to fix and left some angry customers behind.

Configuration

You can use configuration files to make sure that the same level and the same folders are always checked. The configuration is written in NEON (`https://ne-on.org/`), a file format that is very similar to YAML; if you can read and write YAML, it will work out just fine.

The basic configuration only contains the level and the folders to be scanned:

```
parameters:
    level: 4
    paths:
        - src
```

It is a good practice to save this configuration in a file named `phpstan.neon` in the `root` folder of your project. That is the location where *PHPStan* expects it to be by default. If you follow this convention, the next time you want to run it, you only need to specify the desired action:

```
$ vendor/bin/phpstan analyse
```

If you used the above example configuration, *PHPStan* will now scan the `src` folder, using all rules from level 0 to level 4.

That is not everything you can configure here. In the next section, we will learn about some additional parameters.

Using PHPStan in legacy projects

If you want to use *PHPStan* in existing projects of a certain age, you will most likely end up with hundreds if not thousands of errors, depending on the chosen level. Of course, you can decide to keep using a lower level; but that also means that the analyzer will miss more bugs, not only existing ones but also in new or modified code.

In an ideal world, you would start with level 0, solve all errors, then continue with level 1, solve all new errors, and so on. This requires a lot of time, though, and, if no automated tests are available, a complete manual test run at the end would be necessary. You probably won't have that much time, so let us see what other options we have.

There are two ways that *PHPStan* can be told to ignore errors: firstly, using *PHPDocs* annotations, and secondly, using a special parameter in the configuration file.

Using PHPDocs annotations

To ignore a line of code, simply add a comment before or on the affected line, using the special @ phpstan-ignore-next-line and @phpstan-ignore-line *PHPDocs* annotations:

```
// @phpstan-ignore-next-line
$exampleClass->foo();
$exampleClass->bar(); // @phpstan-ignore-line
```

Both lines of code will not be scanned for errors anymore. It is up to you to choose the way you prefer. It is not possible to ignore bigger code blocks or even entire functions or classes, though (unless you want to add a comment to every line, that is).

Using ignoreErrors parameters

The *PHPDocs* annotations are perfect for quick fixes in only a few locations, but you will need to touch many files if you wish to ignore numerous errors. Using the ignoreErrors parameter in the configuration file is not very comfortable, though, as you have to write a regular expression for every error you would like to ignore.

The following example will explain how it works. Let's assume we keep getting an error as follows:

```
Method OrderPosition::getGrossPrice() has no return type
specified.
```

Although theoretically, this would be easy to fix, the team decides against adding a type hint so as not to risk any side effects. The OrderPosition class is awfully written and not covered with tests, yet still works as expected. Since it will be replaced soon anyway, we are not willing to take the risk and touch it.

To ignore this error, we need to add the `ignoreErrors` parameter to our `phpstan.neon` configuration file:

```
parameters:
    level: 6
    paths:
        - src
    ignoreErrors:
        - '#^Method OrderPosition\:\:getGrossPrice\(\) has no
return type specified\.$#'
```

Instead of defining a rule or ruleset to ignore, we need to provide a regular expression here that matches the message of the error that should be ignored.

> **Tip**
>
> Writing regular expressions can be challenging. Luckily, the *PHPStan* website offers a very useful little tool to generate the necessary `phpstan.neon` part from the error message: `https://phpstan.org/user-guide/ignoring-errors#generate-an-ignoreerrors-entry`.

Upon the next run, the error will no longer be displayed regardless of where it occurs, as it matches the regular expression here.

PHPStan does not inform you about the fact that errors are ignored. Do not forget to fix them at some point! However, if you improve your code further over time, *PHPStan* will let you know when errors that are set to be ignored are no longer matched. You can safely remove them from the list then.

If you want to ignore certain errors completely, but just in one or more files or paths, you can do so by using a slightly different notation:

```
ignoreErrors:
    -
        message: '#^Method
            OrderPosition\:\:getGrossPrice\(\) has no return
            type specified\.$#'
        path: src/OrderPosition.php
```

The path needs to be relative to the location of the `phpstan.neon` configuration file. When given, the error will only be ignored if it occurs in `OrderPosition.php`.

Baseline

As we just saw in the previous section, adding errors you want to be ignored manually to your configuration file is a cumbersome task. But there is an easier way: similar to *PHPMD*, it is possible to automatically add all the current errors to the list of ignored errors at once by executing the following command with the `--generate-baseline` option:

```
$ vendor/bin/phpstan analyse --generate-baseline
```

The newly generated file, `phpstan-baseline.neon`, is in the same directory as the configuration file. PHPStan will not make use of it automatically, though. You have to include it manually in the `phpstan.neon` file as follows:

```
includes:
    - phpstan-baseline.neon

parameters:
    ...
```

The next time you run PHPStan now, any previously reported errors should not be reported anymore.

Internally, the baseline file is nothing more than an automatically created list of the `ignoreErrors` parameters. Feel free to modify it to your needs. You can always regenerate it by executing `phpstan` using the `--generate-baseline` option again.

Extensions

It is possible to extend the functionality of *PHPStan*. The vivid community has already created a respectable number of useful extensions. For example, frameworks such as Symfony, Laminas, or Laravel often make use of magic methods (such as `__get()` and `__set()`), which cannot be analyzed automatically. There are extensions for these frameworks that provide the necessary information to *PHPStan*.

While we cannot cover these extensions in this book, we encourage you to check out the extension library: `https://phpstan.org/user-guide/extension-library`. There are also extensions for PHPUnit, phpspec, and WordPress.

A recap of PHPStan

PHPStan is a powerful tool. We cannot cover all its functionality in just a few pages but we have given you a good idea of how to start using it. Once you are familiar with its basic usage, check out `https://phpstan.org` to learn more!

Psalm: A PHP static analysis linting machine

The next and last static code analyzer we want to introduce is *Psalm*. It will check our code base for so-called issues and report any violations. Furthermore, it can resolve some of these issues automatically. So, let us have a closer look.

Installation and usage

Once again, installing *Psalm* with Composer is just a matter of a few keystrokes:

```
$ composer require --dev vimeo/psalm
```

It is available as a `phar` file as well.

After installation, we cannot just start, though – rather, we need to set up a configuration file for the current project first. We can use the comfortable `--init` option to create it:

```
$ vendor/bin/psalm --init
```

This command will write a configuration file called `psalm.xml` in the current directory, which should be the project root. During its creation, *Psalm* checks whether it can find any PHP code and decides which error level is suitable, to begin with. Running *Psalm* doesn't require any more options:

```
$ vendor/bin/psalm
```

Configuration

The configuration file was already created during the installation process and could, for example, look similar to this:

```
<?xml version="1.0"?>
<psalm
    errorLevel="7"
    resolveFromConfigFile="true"
    xmlns:xsi=http://www.w3.org/2001/XMLSchema-instance
    xmlns=https://getpsalm.org/schema/config
    xsi:schemaLocation=https://getpsalm.org/schema/config
vendor/vimeo/psalm/config.xsd
>
    <projectFiles>
        <directory name="src" />
        <ignoreFiles>
            <directory name="vendor" />
```

```
        </ignoreFiles>
    </projectFiles>
</psalm>
```

Let us have a look at the attributes of the `<psalm>` node. You do not need to worry about the schema- and namespace-related information, only about the following two things:

- `errorLevel`: The levels go from 8 (basic checks) to 1 (very strict). In other words, the lower the level, the more rules will be applied.

- `resolveFromConfigFile`: Setting this to `true` lets *Psalm* resolve all relative paths (such as `src` and `vendor`) from the location of the configuration file – so usually, from the project root.

> **Psalm documentation**
>
> Psalm offers many more configuration options that we cannot cover in this book. As always, we recommend checking the documentation (`https://psalm.dev/docs`) to learn more about this tool.

Inside the `<psalm>` node, you will find more settings. In the previous example, *Psalm* is told to only scan the `src` folder and ignore all the files in the `vendor` folder. Ignoring the `vendor` folder is important, as we don't want to scan any third-party code.

Using Psalm in legacy projects

We will now have a look at how we can adjust *Psalm* to deal with existing projects better. As with the previous tools, there are basically two ways to ignore issues: using the configuration file or docblock annotations.

There are three code issue levels: `info`, `error`, and `suppress`. While `info` will just print info messages if minor issues have been found, issues that are at the level of an `error` type, on the other hand, require you to get active. An issue of the `suppress` type will not be shown at all.

> **Continuous Integration**
>
> The difference between `info` and `error` becomes more important when building a Continuous Integration pipeline. `info` issues would let the build pass, while `error` issues would break it. We will have a closer look at this topic later.

Docblock suppression

The `@psalm-suppress` annotation can be used either in a function docblock or a comment for the next line. The `Vat` class from the previous examples could look as follows:

```php
class Vat
{
    private float $vat = 0.19;


    /**
     * @psalm-suppress InvalidReturnType
     */
    public function getVat(): int
    {
        /**
         * @psalm-suppress InvalidReturnStatement
         */
        return $this->vat;
    }
}
```

Configuration file suppression

If we want to suppress issues, we need to configure issueHandler for them, where we can set the type to suppress manually. This is done in the configuration file by adding an <issueHandler> node inside the <psalm> node:

```xml
<issueHandlers>
    <InvalidReturnType errorLevel="suppress" />
    <InvalidReturnStatement errorLevel="suppress" />
</issueHandlers>
```

The preceding configuration would suppress all the InvalidReturnType and InvalidReturnStatement issues in the whole project. We can make this a bit more specific, though:

```xml
<issueHandlers>
    <InvalidReturnType>
        <errorLevel type="suppress">
            <file name="Vat.php" />
        </errorLevel>
    </InvalidReturnType>
    <InvalidReturnStatement>
        <errorLevel type="suppress">
            <dir name="src/Vat" />
```

```
        </errorLevel>
    </InvalidReturnStatement>
</issueHandlers>
```

In the documentation (`https://psalm.dev/docs/running_psalm/dealing_with_code_issues/`), you will find even more ways to suppress issues – for example, by the variable name.

Baseline

As with the previous static code analyzers we discussed, *Psalm* also provides a feature to generate a baseline file, which will include all the current errors so that they will be ignored during the next run. Please note that the baseline feature only works for `error` issues, but not `info` issues. Let us create the file first:

```
$ vendor/bin/psalm --set-baseline=psalm-baseline.xml
```

Psalm has no default name for this file, so you need to pass it as an option to the command:

```
$ vendor/bin/psalm --use-baseline=psalm-baseline.xml
```

You can also add it as an additional attribute to the `<psalm>` node in the configuration file:

```
<psalm
    ...
    errorBaseline="./psalm-baseline.xml"
>
```

Finally, you can update the baseline file – for example, after you have made some improvements to the code:

```
$ vendor/bin/psalm --update-baseline
```

Fixing issues automatically

Psalm will not only find the issue but it can also fix many of them automatically. It will let you know when this is the case and you can use the `--alter` option:

```
Psalm can automatically fix 1 issues.
Run Psalm again with
--alter --issues=InvalidReturnType --dry-run
to see what it can fix.
```

Let's execute the command as *Psalm* suggests:

```
$ vendor/bin/psalm --alter --issues=InvalidReturnType --dry-run
```

The `--dry-run` option tells *Psalm* to only show you what it would change as `diff`, but not to apply the changes. This way, you can check whether the change is correct:

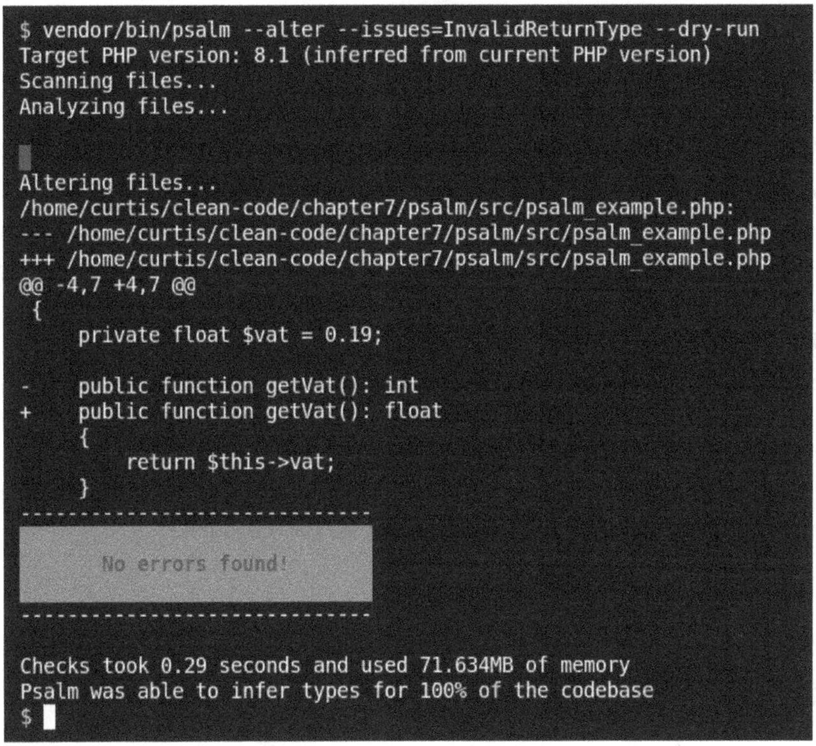

```
$ vendor/bin/psalm --alter --issues=InvalidReturnType --dry-run
Target PHP version: 8.1 (inferred from current PHP version)
Scanning files...
Analyzing files...

Altering files...
/home/curtis/clean-code/chapter7/psalm/src/psalm_example.php:
--- /home/curtis/clean-code/chapter7/psalm/src/psalm_example.php
+++ /home/curtis/clean-code/chapter7/psalm/src/psalm_example.php
@@ -4,7 +4,7 @@
 {
     private float $vat = 0.19;

-    public function getVat(): int
+    public function getVat(): float
     {
         return $this->vat;
     }
--------------------------------
        No errors found!
--------------------------------

Checks took 0.29 seconds and used 71.634MB of memory
Psalm was able to infer types for 100% of the codebase
$
```

Fig 7.4: Psalm showing proposed changes

If you remove the `--dry-run` option, the changes will be applied.

A recap on Psalm

Psalm is a standard tool in the clean coder's toolkit for good reason. It is fast, easy to use, and powerful. Additionally, the code manipulation feature will save you a lot of time. Of course, there are numerous similarities with *PHPStan*, but often enough, you will find both tools working together on the same code base without problems. At least, you should consider giving it a try.

IDE extensions

The tools we looked at so far share something in common: they need to be applied to our code after we have written it. Of course, this is much better than nothing, but wouldn't it be great if the tools gave us their feedback immediately at the time that we wrote the code?

That is what many other developers thought as well, so they created extensions for the most popular IDEs, which are currently **Visual Studio Code** (**VS Code**) and **PhpStorm**:

- *PhpStorm* is an established, commercial IDE from JetBrains with several PHP-specific tools, checks, and built-in integrations for many of the code quality tools we discussed in this chapter. There are many useful extensions available for it as well. You can try it out for 30 days for free.

- *VS Code* is a highly flexible code editor from Microsoft with tons of third-party (partly commercial) extensions that can turn these tools into an IDE for virtually every relevant programming language today. Because the code editor itself is free, is it becoming more and more popular.

> **Alternative PHP IDEs**
>
> *PhpStorm* and *VS Code* are not the only IDEs that exist for PHP. Other alternatives are *NetBeans* (`https://netbeans.apache.org`), *Eclipse PDT* (`https://www.eclipse.org`), or *CodeLobster* (`https://www.codelobster.com`).

In this section, we will introduce you to three extensions for these two IDEs:

- **PHP Inspections** (EA Extended) for PhpStorm
- **Intelephense** for VS Code

> **Code quality tool integration in PhpStorm**
>
> PhpStorm offers seamless integration for the following tools that we have discussed: *PHP CS Fixer*, PHPMD, PHPStan, and Psalm. More information can be found here: `https://www.jetbrains.com/help/phpstorm/php-code-quality-tools.html`.

PHP Inspections (EA Extended)

This plugin (`https://github.com/kalessil/phpinspectionsea`) is for PhpStorm. It will add even more types of inspections to the pool of already existing ones, covering topics such as code style, architecture, or possible bugs.

> **IDE Inspections**
>
> Modern IDEs are already equipped with a lot of useful code checks. In PHPStorm, they are called *Inspections*. Some are already enabled by default – more can be activated manually (`https://www.jetbrains.com/help/phpstorm/code-inspection.html#access-inspections-and-settings`). For VS Code, you need to install an extension first. Check out the documentation (`https://code.visualstudio.com/docs/languages/php`) for more information.

Installation

As with every PhpStorm plugin, the installation is done via the **File** -> **Settings** -> **Plugins** dialog. You will find detailed information on how to install a plugin on the vendor's website (`https://www.jetbrains.com/help/phpstorm/managing-plugins.html`). Simply search for `EA Extended`. Please note that there is a second version of this plugin, EA Ultimate, which you have to pay for. We will not cover it in this book.

After installation, not all the inspections are immediately active. Let us have a look at the PhpStorm inspections configuration, as shown in *Figure 7.4*:

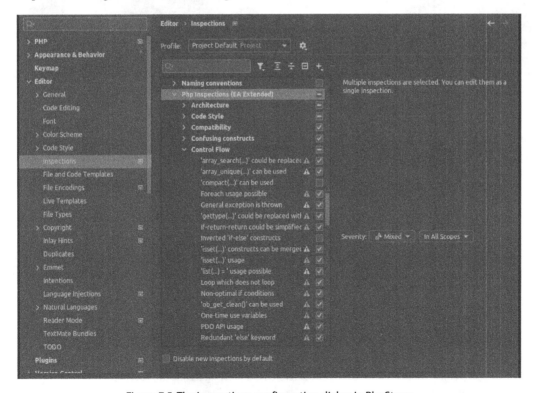

Figure 7.5: The Inspections configuration dialog in PhpStorm

All the inspections of this plugin can be found in the **Php Inspections (EA Extended)** section. The inspections that are not active by default can easily be activated by checking the checkbox next to them. We recommend reading the documentation (`https://github.com/kalessil/phpinspectionsea/tree/master/docs`) before activating any further inspections – otherwise, you might end up with too many rules. You can revisit them later.

Usage

PHP Inspections (EA Extended) not only warns you about problems but often also offers so-called Quick-Fixes, which let the IDE do the work for you. Here, you will find an example. Note the highlighted `if` clause on line 7:

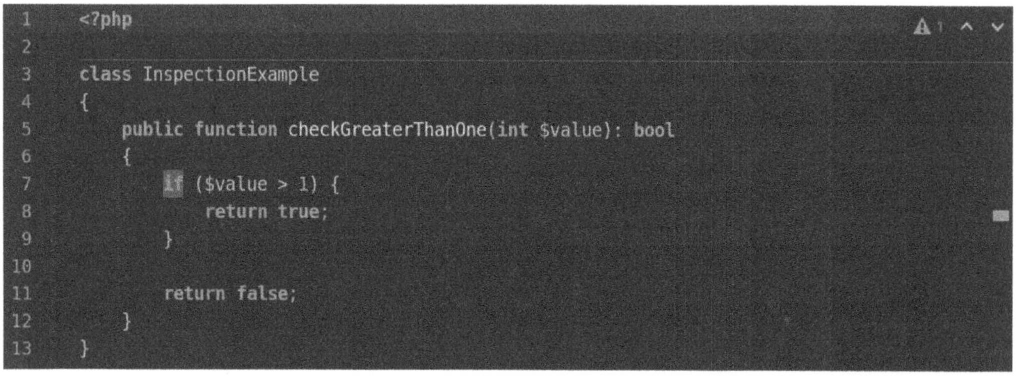

```php
1   <?php
2
3   class InspectionExample
4   {
5       public function checkGreaterThanOne(int $value): bool
6       {
7           if ($value > 1) {
8               return true;
9           }
10
11          return false;
12      }
13  }
```

Figure 7.6: Example code with an issue found by PHP Inspections (EA Extended)

When you hover your mouse pointer over the highlighted area, PhpStorm will show a pop-up window with further instructions about the suggested improvement:

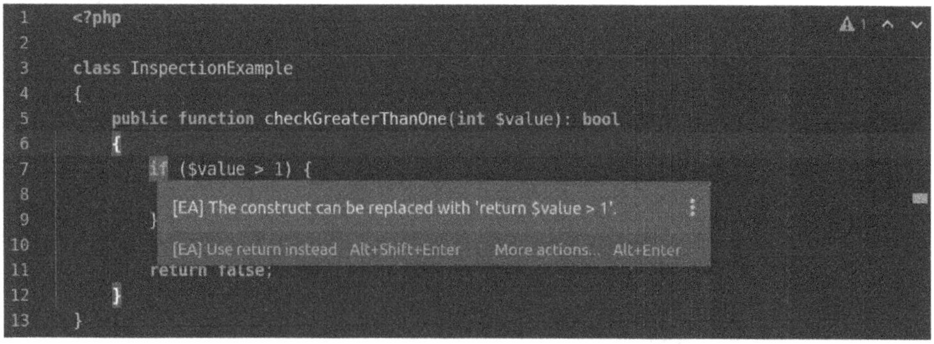

```php
1   <?php
2
3   class InspectionExample
4   {
5       public function checkGreaterThanOne(int $value): bool
6       {
7           if ($value > 1) {
8               [EA] The construct can be replaced with 'return $value > 1'.
9           }
10              [EA] Use return instead  Alt+Shift+Enter    More actions...  Alt+Enter
11          return false;
12      }
13  }
```

Figure 7.7: PHP Inspections (EA Extended) suggesting a code improvement

You can choose to fix the issue directly by pressing *Alt + Shift + Enter* at the same time, or you can click on the highlighted area to show the Quick-Fix bubble. If you click on the bubble, you will see a menu with some more options. You can also invoke the following dialog by pressing *Alt + Enter*:

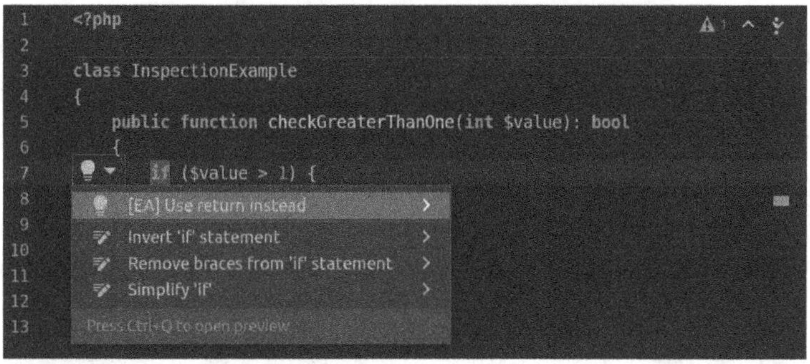

Figure 7.8: The Quick-Fix options menu

PhpStorm offers you several fixes now. The first one, marked with `[EA]`, is a suggestion by the plugin. Another click will apply the fix:

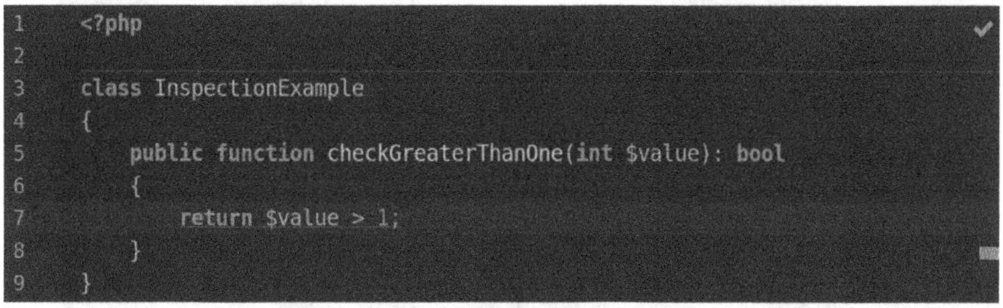

Figure 7.9: The code after applying a Quick-Fix

That's it! Within just a few seconds, you made your code shorter and easier to read. **PHP Inspections (EA Extended)** is a great addition to PhpStorm, as it offers sensible inspections and integrates them seamlessly. If you are using this IDE, you should not hesitate to install it.

> **Inspections when working in a team**
>
> These inspections are a great way to improve your code and educate yourself on best practices. However, there is a huge drawback: how do you ensure that every developer working on your project has the same inspections activated? We will cover this topic in *Working in a Team*.

Intelephense

The second extension we want to introduce is *Intelephense* for VS Code. It is the most frequently downloaded PHP extension for this editor and provides a lot of functionality (such as code completion and formatting), which turns VS Code into a fully capable PHP IDE. There is also a commercial, premium version of this extension that offers even more functionality. To install it, please follow the instructions on the Marketplace website for this plugin (`https://marketplace.visualstudio.com/items?itemName=bmewburn.vscode-intelephense-client`).

Intelephense does not have the range of functionality that a full-grown, commercial IDE would offer by any means, yet for a free service, it is a perfect choice. It offers so-called Diagnostics (which are similar to Inspections in PhpStorm) that can be configured in the plugin settings screen, as shown in *Figure 7.9*:

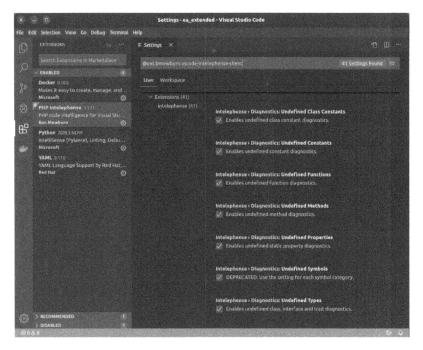

Figure 7.10: The Intelephense settings screen

Usage

The following figure shows Diagnostics in Intelephense in action:

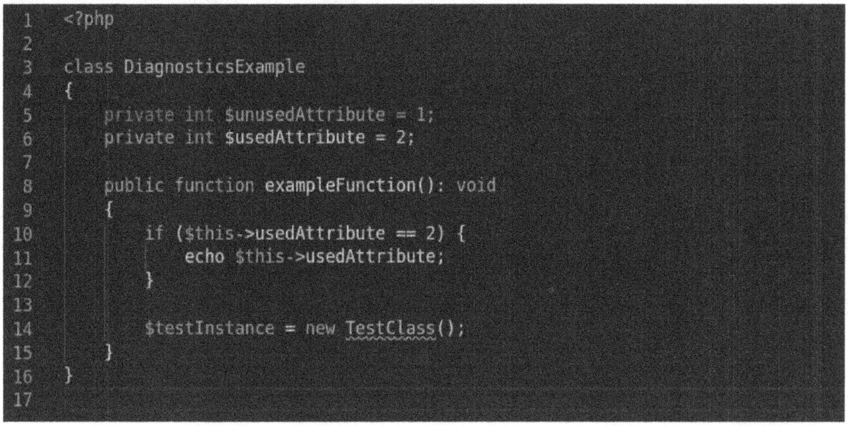

```php
1   <?php
2
3   class DiagnosticsExample
4   {
5       private int $unusedAttribute = 1;
6       private int $usedAttribute = 2;
7
8       public function exampleFunction(): void
9       {
10          if ($this->usedAttribute == 2) {
11              echo $this->usedAttribute;
12          }
13
14          $testInstance = new TestClass();
15      }
16  }
17
```

Figure 7.11: A sample class showing how Intelephense highlights issues

Two things can be seen here. Firstly, and more obviously, is the red line underneath `TestClass`. Hovering the mouse pointer over `TestClass` will show a pop-up window with an explanation: **Undefined type TestClass**. This makes sense since this class does not exist.

Secondly, and more subtly, you will notice that `$ununsedAttribute` and `$testInstance` have a slightly darker color than the other variables. This indicates another issue, which can be revealed by hovering the mouse over one of the variables:

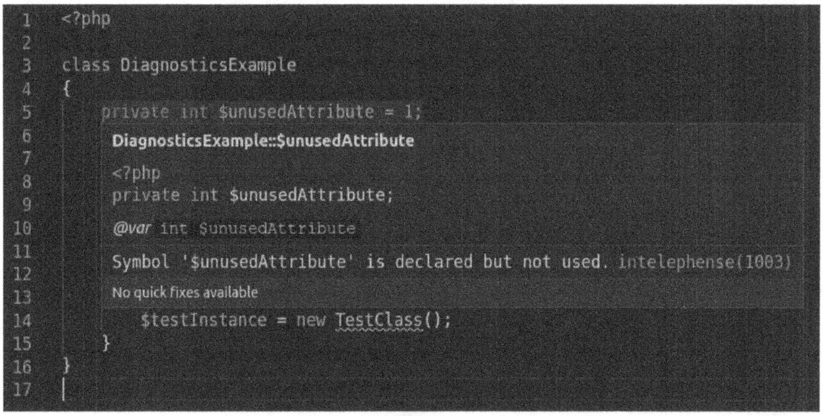

```php
1   <?php
2
3   class DiagnosticsExample
4   {
5       private int $unusedAttribute = 1;

        DiagnosticsExample::$unusedAttribute

        <?php
        private int $unusedAttribute;

        @var int $unusedAttribute

        Symbol '$unusedAttribute' is declared but not used. intelephense(1003)

        No quick fixes available
14          $testInstance = new TestClass();
15      }
16  }
17  [
```

Figure 7.12: An info popup in Intelephense

The popup tells us that `$unsuserAttribute` is not used elsewhere in the code. The same applies to `$testInstance` as well.

Although it provides some basic issue detection rules and code formatting, it clearly can be said that, at the time of writing, the focus of this plugin is not on clean code. However, given the fact that VS Code and this plugin are freely available, you already have a decent PHP IDE on hand to start coding.

Code quality tool integration in VS Code

As in PhpStorm, it is possible to integrate some common code quality tools into VS Code using plugins, such as for PHPStan (`https://marketplace.visualstudio.com/items?itemName=calsmurf2904.vscode-phpstan`), *PHP CS Fixer* (`https://marketplace.visualstudio.com/items?itemName=junstyle.php-cs-fixer`), and PHPMD (`https://marketplace.visualstudio.com/items?itemName=ecodes.vscode-phpmd`). So, if you want to code with VS Code, be sure to check Marketplace for new plugins every now and then.

Summary

In this chapter, we learned about state-of-the-art tools to assist you in creating high-quality PHP code. They will help you spot issues early in the **Software Development Life Cycle** (**SDLC**), which saves you vast amounts of time. The PHP community is still vivid and very productive, and we were not able to cover all the fantastic software that exists out there in this book. However, with the tools we introduced in this chapter, you are now well equipped for your journey towards clean code.

In the next chapter, you will learn about how to evaluate code quality by using the established metrics and, of course, the necessary tools to gather them. See you there!

Further reading

If you want to try out even more code quality tools, consider the following projects:

- *Exakat* (`https://www.exakat.io`) – A tool that also covers security issues and performance, for example. It can fix issues automatically, too.

- *Phan* (`https://github.com/phan/phan`) – A static code analyzer that you can try out immediately in your browser

- *PHP Insights* (`https://phpinsights.com/`) – Another analyzer, yet with easy-to-use metrics in terms of the code, architecture, complexity, and style

8

Code Quality Metrics

Wouldn't it be great if we could measure the quality of our software? Software developers often want to improve their software again and again – but maybe it is already "good enough." How do we know when it reaches a good state?

Software quality metrics were introduced by smart people in the early days of programming. In the 1970s, they thought about this topic and produced the ideas that are still in use today. Of course, we want to benefit from this knowledge and apply it to our own projects.

In this chapter, we are going to cover the following topics:

- Introducing code quality metrics
- Gathering metrics in PHP
- The pros and cons of using metrics

Technical requirements

If you have gone through the previous chapter and tried out all the tools, you already have everything that you need for this chapter installed. If not, please make sure to do so before you run the upcoming examples.

The code files for this chapter can be found here: `https://github.com/PacktPublishing/Clean-Code-in-PHP`

Introducing code quality metrics

In this section, you will learn about how to measure the quality of software in general. We will look at some of the most used metrics in the PHP world and explain what they can tell you about your code, how to gather them, and when they are useful or not.

Aspects of software quality

Before we dive into the numbers, we need to clarify one important thing first: what does software quality actually mean? Surely, everybody has a certain understanding of quality, but it might be hard to put this in words. Luckily, there are already existing models, such as the **FURPS** model, which was developed at Hewlett-Packard back in the 1980s. The acronym stands for the following:

- **Functionality**: Is the software capable of dealing with a wide variety of use cases? Has it been developed with security in mind?

- **Usability**: How good is the user experience? Is it documented and easy to understand?

- **Reliability**: Is the software available at all times? How probable are crashes, or errors, that might affect the output?

- **Performance**: Indicates the speed of the software. Does it make efficient use of the available resources? Does it scale well?

- **Supportability**: Can the software be tested and maintained well? Is it easy to install and can it be translated (localized) into other languages?

Further quality aspects include, among others, accessibility and legal conformity. As you can see, this model covers more aspects such as user experience and documentation than we as PHP developers will typically work on. That is why we can look at software quality from two different viewpoints: external and internal quality. Let us have a closer look at what that means:

- **External quality**: Outward, or user-facing, aspects are part of the external quality of software. This covers a lot of the aspects we introduced previously. What they have in common is that they can be measured without touching or analyzing the code itself – think of performance testing tools that measure the response time of a request or end-to-end tests that emulate a user by automatically executing tests on the application.

- **Internal quality**: As software developers, we usually care more about the internal quality of software. Is the code easy to read and understand? Can you extend it easily? Can we write tests for it? While users will never see the code, or are not concerned about its testability, it does affect them indirectly: code of high quality contains lesser bugs and is also often (but not always) faster and more efficient. It is also known to be easier to extend and maintain. Typically, these aspects can be checked using automated unit tests or code analyzers.

In this book, we focus on the internal code quality. That is why we speak about code quality in particular and don't use the broader term, software quality.

Code quality metrics

Now that we have a better understanding of what code quality means, let us now have a look at what code quality metrics we want to talk about in this section:

- Lines of code

- The cyclomatic complexity

- The NPath complexity

- Halstead metrics

- The Change Risk Anti-Patterns index

- The maintainability index

The lines of code

Counting the **Lines of Code (LOC)** in a project is not a quality metric. However, it is a useful tool to grasp the size of a project – for example, when you start working on it. Furthermore, as we will see, it is used by other metrics as a base for their calculations. It is also helpful to have an idea about how many lines of code you are dealing with – for example, when you need to estimate refactoring efforts for certain classes.

That is why we want to have a closer look at it now. First, we can differentiate the LOC further:

- **LOC**: LOC simply counts all lines of code, including comments and blank lines.

- **Comment Lines of Code (CLOC)**: This metric tells you how many lines of your code are comments. It can be an indicator of how well the source code is commented on. However, as we know, comments tend to rot (i.e., they get outdated quickly and are often more harmful than they are useful), so there is no percentage or any other rule of thumb we can recommend. Still, it is interesting to know.

- **Non-Comment Lines of Code (NCLOC)**: If you want to compare the size of one project with another, leaving out the comments will give you a better picture of how much real code you need to deal with.

- **Logical Lines of Code (LLOC)**: For this metric, it is assumed that every statement equals one line of code. The following code snippet illustrates how it is supposed to work. Consider the following line of code:

```
while($i < 5) { echo "test"; /* Increment by one */
    $i++; }
```

Here, the LOC would be 1. Since we have three executable statements in this line, LLOC would count this as 3, as the code can also be written with each statement in one line:

```
while($i < 5) {
    echo "test";
    /* Increment by one */
    $i++;
}
```

In the preceding example, we highlighted the executable statements. Comments, empty lines, and syntactical elements such as brackets are not executable statements – that is why the full-line comment and the closing brace at the end of the loop are not counted as a logical line.

The cyclomatic complexity

Instead of just counting the lines of code, we can also measure the complexity of the code – for example, by counting the number of execution paths within a function. A common metric for this is the **Cyclomatic Complexity** (**CC**). It was introduced in the late 1970s but is nevertheless still useful. The idea behind the cryptic name is simple: we count the number of decision points, which are if, while, for, and case statements. Additionally, the function entry counts as one statement as well.

The following example illustrates how the metric works:

```
// first decision point
function someExample($a, $b)
{
    // second decision point
    if ($a < $b) {
        echo "1";
    } else {
        echo "2";
    }

    // third decision point
    if ($a > $b) {
        echo "3";
    } else {
        echo "4";
    }
}
```

The CC for the preceding code snippet would be 3: the function entry counts as the first decision path and both `if` statements count as one decision path each as well. However, both `else` statements are not taken into account by definition, as they are part of the `if` clauses. This metric is especially useful to quickly assess the complexity of code that you do not know yet. It is often used to check a single function, but can also be applied to classes or even a whole application. If you have a function with a high CC, consider splitting it into several smaller functions to reduce the value.

The NPath complexity

A second metric of code complexity is the **NPath complexity**. The basic idea is similar to the CC, as it counts the decision paths of a function too. However, it counts *all* possible decision paths and not just the four statements (`if`, `while`, `for`, and `case`) that are defined for the CC. Furthermore, the function entry point is not counted as a decision path for this metric.

Looking at the above example, the NPath complexity would be 4, because we have 2 * 2 possible paths through the function: both `if` statements, as well as both `else` statements. So, all four `echo` statements are considered decision paths. As mentioned previously, the function call itself is not considered. Now, if we added another `if` statement, the NPath complexity would increase to 8. This is because we would then have 2 * 2 * 2 possible paths. In other words, the metric grows exponentially, so it can rapidly become quite high.

The NPath complexity depicts the actual effort to test a function better than the CC, as it tells us directly how many possible outcomes of the function we would need to test to achieve 100% test coverage.

Halstead metrics

Maurice Halstead introduced a set of eight metrics in the late 1970s, which are still in use today and are known as the **Halstead metrics**. They are based solely on the distinct and total number of operators (e.g., `==`, `!=`, and `&&`) and operands (e.g., function names, variables, and constants), but as you will see, they already tell you a lot about the inspected code.

We do not need to know exactly how these metrics work. If you are interested, you can find out more about these metrics here: `https://www.verifysoft.com/en_halstead_metrics.html`. However, you should have an idea of what Halstead metrics there are:

- *Length*: Calculating the sum of the total number of operators and operands tells us how much code we must deal

- *Vocabulary*: The sum of the number of unique operators and operands already indicates the complexity of the code

- *Volume*: Describes the information content of the code based on the length and vocabulary

- *Difficulty*: Indicates the error proneness (i.e., how likely it is to introduce bugs)

- *Level*: Inverts the difficulty – as in, the higher the level, the less error-prone it is

- *Effort*: The effort that is necessary to understand the code

- *Time*: Tells us how long it roughly took to implement it

- *Bugs*: Estimates the number of bugs that the code contains

These values will give you a rough indication of what type of code you are dealing with. Is it easy to understand? How much time was spent developing it? How many bugs can be expected? However, without comparing these values with results from other applications, they will not help you that much.

The Change Risk Anti-Patterns index

Another especially useful metric is the **Change Risk Anti-Patterns** (**CRAP**) index. It uses the CC and the code coverage of the code under consideration.

> **Code coverage**
>
> You have probably heard the term code coverage a lot already. It is a metric that is used in context with automated tests and describes the number of lines of code (stated in percent of the total number of lines) that unit tests have been written for. We will discuss this metric and its prerequisites again later in the book when we are dealing with this topic in more detail.

The combination of these two metrics is quite useful. Code that is not overly complex and has high test coverage is far more likely to be bug-free and maintainable than code that is complex and where there are not many tests for it.

The maintainability index

As the last metric in this section, we will look at the **maintainability index**. It will provide you with just one value that indicates the maintainability of the inspected code, or, in other words, it tells you how easy it will be to change it without introducing new bugs. Two things make this metric particularly interesting for us.

Firstly, it is based on the aforementioned metrics and uses the LOC, Halstead metrics, and the CC to calculate the index. Yet again, we do not really need to know the exact formula. If you are interested, you can look it up here: `https://www.verifysoft.com/en_maintainability.html`.

Secondly, this metric will return a value that you can use to assess the code quality directly:

- 85 and more: Good maintainability

- 65 to 85: Moderate maintainability

- 65 and below: Bad maintainability

With this metric, you need no other code to compare it to. That is why it is particularly useful to quickly assess the code quality.

In this section, we have gone through a lot of theory. Excellent job so far – you will not regret learning about it for sure because, in the next section, we will show you how to gather these metrics using even more PHP tools.

Gathering metrics in PHP

In this section, we want to have a look at the tools there are in the PHP world to gather code quality metrics. As you will see shortly, these metrics are not just numbers – they will allow you to make educated guesses about how much effort it will take to refactor code. They will also help you to identify the parts of the code that will require the most attention.

Again, we have curated a selection of tools for you:

- `phploc`
- PHP Depend
- PhpMetrics

phploc

As we learned in the previous section, the abbreviation LOC stands for lines of code, so the name already reveals the main purpose of this tool. Being a basic metric, it already tells us quite a few things about a code base. `phploc` also provides further metrics, such as the CC, so it is worth having a closer look at it.

Installation and usage

The author of this tool, Sebastian Bergmann, is well known for `phpunit`, the de facto standard for automated tests in the PHP world. He recommends not installing it using Composer but using `phar` directly. We will discuss the pros and cons of this approach in the next chapter. For now, let us just follow the author's advice and download `phar` directly:

```
$ wget https://phar.phpunit.de/phploc.phar
```

This will download the latest version of `phploc` into the current directory. After downloading it, we can directly use it to scan a project:

```
$ php phploc.phar src
```

> **Scanning single files**
>
> Although `phploc` is meant to be used on whole projects, it is also possible to specify a single file to scan. While the average measures make no sense because they are meant to be used on a whole project, it is still useful if you need to find out the LOC metrics or the CC for a class.

The preceding command will scan the `src` folder with all its subfolders and gather information about it, which will be presented directly on the command line as shown in *Figure 8.1*:

```
$ php phploc.phar src
phploc 7.0.2 by Sebastian Bergmann.

Directories                                    8
Files                                         47

Size
  Lines of Code (LOC)                       3349
  Comment Lines of Code (CLOC)               199 (5.94%)
  Non-Comment Lines of Code (NCLOC)         3150 (94.06%)
  Logical Lines of Code (LLOC)               897 (26.78%)
    Classes                                  868 (96.77%)
      Average Class Length                    18
        Minimum Class Length                   0
        Maximum Class Length                 117
      Average Method Length                    3
        Minimum Method Length                  0
        Maximum Method Length                 40
      Average Methods Per Class                3
        Minimum Methods Per Class              1
        Maximum Methods Per Class             23
    Functions                                 29 (3.23%)
      Average Function Length                  1
    Not in classes or functions                0 (0.00%)

Cyclomatic Complexity
  Average Complexity per LLOC               0.22
  Average Complexity per Class              4.89
    Minimum Class Complexity                1.00
    Maximum Class Complexity               51.00
  Average Complexity per Method             2.10
    Minimum Method Complexity               1.00
    Maximum Method Complexity              26.00

Dependencies
  Global Accesses                              0
    Global Constants                           0 (0.00%)
    Global Variables                           0 (0.00%)
    Super-Global Variables                     0 (0.00%)
  Attribute Accesses                         805
    Non-Static                               805 (100.00%)
    Static                                     0 (0.00%)
  Method Calls                               364
    Non-Static                               349 (95.88%)
    Static                                    15 (4.12%)
```

Figure 8.1: An example output of phploc (an excerpt)

That is quite a lot more information than just the LOC. The information is divided into the following categories:

- **Size**: Obviously, the main reason for this tool to exist is to measure the size of a project by counting the number of lines of code, using the several ways of counting we introduced in the previous section. The focus lies on LLOC and you will get the averages for this metric per class, class method, and function.

- **CC**: phploc will calculate the average CC values per LLOC, classes, and methods.

- **Dependencies**: This section tells you how many accesses to the global state have been made and how many attributes and methods are being accessed statically. Both global and static access are considered as practices and should be avoided, so these numbers give you greater hints about the code quality.

- **Structure**: In the last output section (which did not fit in the preceding screenshot), phploc returns more details on the code structure. There are no clear rules on how to interpret them; however, you can draw some conclusions from them. For example, see the following:

  - Regarding the overall code size, how many namespaces are used? A large code base with only a few namespaces indicates that the project is not well structured.

  - Are interfaces used and how many compared to the project size? The usage of interfaces increases the interchangeability of classes and indicates well-structured code.

This is all we need to know about the functionality of phploc for now. It is a simple-to-use yet helpful tool that helps you get a grasp on the overall code quality and structure of a project quickly and should therefore be part of your toolkit. It does not tell you how to interpret the numbers, though, which requires some experience.

PHP Depend

If there was a prize for the most metrics combined in one tool, then it would surely go to **PHP Depend** (**PDepend**). It covers all the metrics we discussed in the previous section, plus many more. However, it is not the most user-friendly tool there is. Plus, the website and the repository documentation are not perfect. Nevertheless, you should check it out.

Installation and usage

As before, this tool can be installed using Composer or downloading phar directly. We will go with the Composer-based installation for now:

```
$ composer require pdepend/pdepend --dev
```

If there were no unpleasant surprises, you can execute it directly:

```
$ vendor/bin/pdepend --summary-xml=pdepend_summary.xml src
```

Here, we already can see that the ancestor of PDepend is JDepend, a Java code quality tool, as the output is written into an XML file. The filename is specified using the --summary-xml option. Furthermore, we must specify the to-be-scanned folder as an argument.

PDepend does output some numbers, though, as can be seen in the following example output:

```
PDepend 2.10.3

Parsing source files:
............................................    47

Calculating Cyclomatic Complexity metrics:
.................    355

Calculating Node Loc metrics:
.............    279

Calculating NPath Complexity metrics:
.................    355

Calculating Inheritance metrics:
.....    101
```

We skipped some lines here. The numbers will only tell you how often each metric has been calculated for the given folder, so the direct output is not particularly helpful. To see the actual metrics, we need to open the XML report. In our case, the file that has been generated is called pdepend_summary.xml.

The XML report is too huge to print in this book, so you best try it out yourself to see it in all its glory. However, we can show you how it is structured:

```
<?xml version="1.0" encoding="UTF-8"?>
<metrics>
  <files>
    <file name="/path/to/Namespace/Classname.php"/>
    <!-- ... -->
  </files>
```

```
  <package name="Namespace">
    <class name="Classname" fqname="Namespace\Classname">
      <file name="/path/to/Namespace/Classname.php"/>
      <method name="methodName"/>
      <!-- ... -->
    </class>
    <!-- ... -->
  </package>
</metrics>
```

The `<metrics>` node represents the directory that was scanned in its entirety. It has the following child nodes:

- `<files>`, which lists all the scanned files using the `<file>` child nodes.
- `<package>`, which lists all the namespaces. Within this node, there are further `<class>` child nodes. For each class, there is a list of `<method>` nodes, one for each method in the class. Finally, the filename of the class is mentioned in another `<file>` node.

Of course, this is not everything that *PDepend* will generate as output. For each node, it will add up dozens of attributes, which contain the names and values of the calculated metrics. This is an example node from a XML report that was generated on the source code of *PDepend* itself:

```
<method name="setConfigurationFile" start="80" end="89"
  ccn="2" ccn2="2" loc="10" cloc="0" eloc="8" lloc="3"
  ncloc="10" npath="2" hnt="15" hnd="21"
  hv="65.884761341681" hd="7.3125" hl="0.13675213675214"
  he="481.78231731105" ht="26.765684295058"
  hb="0.020485472371812" hi="9.0098818928795"
  mi="67.295865328327"/>
```

You should be able to recognize some metrics such as `lloc` (LOC) or `ccn` (CC Number) already. For the others, you will find explanations, or at least the long names, for the abbreviations in the *XML* report in the online documentation: `https://pdepend.org/documentation/software-metrics/index.html`.

Further options

PDepend has two options you should know about:

- --exclude: This will exclude a namespace (or package, in this terminology) from the scans. You can use multiple namespaces, separated by commas. Make sure to add quotes around the namespace(s):

```
$ vendor/bin/pdepend --summary-xml=pdepend_summary.xml
--exclude="Your\Namespace,Another\Namespace" src
```

- --ignore: Allows you to ignore one or more folders. Again, don't forget the quotes:

```
$ vendor/bin/pdepend --summary-xml=pdepend_summary.xml
--ignore="path/to/folder,path/to/other/folder" src
```

It can also generate images in SVG format with further information. We will not cover them in this book, though, as there is a better tool for this, which you will find in the next section.

PDepend is powerful, but at the same time difficult to oversee. The generated output is hard to read and, once the project has become a bit bigger, becomes unusable unless you use other tools to parse the XML file. However, you may need the advanced metrics it provides one day, or you may work on a project where it is already in use. So, at least you are prepared now.

PhpMetrics

Up until now, the world of PHP quality metrics was text-based only. This will change now, as we will now have a look at *PhpMetrics*, which will generate reports that are much better suited for the human eye and are even interactive.

Installation and usage

Let us add *PhpMetrics* to your project using Composer:

```
$ composer require phpmetrics/phpmetrics --dev
```

After all the files have been downloaded, you can immediately start generating your first report:

```
$ vendor/bin/phpmetrics --report-html=phpmetrics_report src
```

The --report-html option specifies the folder in which the report will be created. You can specify more than one folder to be scanned by providing them as a comma-separated list. For our example, however, we will just use the src folder.

As a result, *PhpMetrics* will list several statistics that will already tell you a bit about the code. *Figure 8.2* shows an excerpt of the output, which might remind you of the output generated by `phploc`:

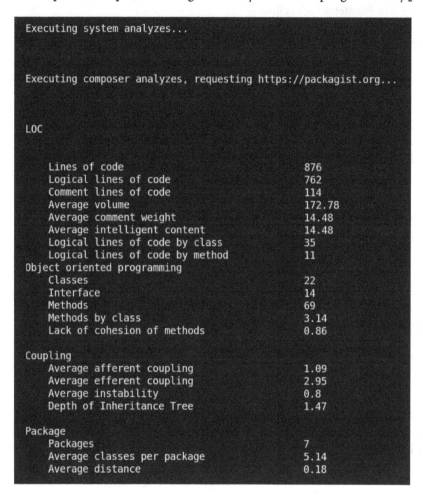

```
Executing system analyzes...


Executing composer analyzes, requesting https://packagist.org...


LOC

        Lines of code                         876
        Logical lines of code                 762
        Comment lines of code                 114
        Average volume                        172.78
        Average comment weight                14.48
        Average intelligent content           14.48
        Logical lines of code by class        35
        Logical lines of code by method       11
Object oriented programming
        Classes                               22
        Interface                             14
        Methods                               69
        Methods by class                      3.14
        Lack of cohesion of methods           0.86

Coupling
        Average afferent coupling             1.09
        Average efferent coupling             2.95
        Average instability                   0.8
        Depth of Inheritance Tree             1.47

Package
        Packages                              7
        Average classes per package           5.14
        Average distance                      0.18
```

Figure 8.2: The PhpMetrics console output (an excerpt)

To open the actual HTML report that has just been generated, simply open the index.html file in that folder in your browser. Before we have a closer look at the generated report, let us see which other useful options *PhpMetrics* offers first:

- --metrics: This option will return a list of the available metrics. It helps decipher abbreviations such as mIwoC.

- --exclude: With this option, you can specify one or more directories to be excluded from scanning.

- --report-[csv|json|summary-json|violations]: Allows you to save the results in different report formats other than HTML – for example, --report-json.

Opening the browser from the command line

If you are using a Linux-based operating system, such as Ubuntu, you can quickly open an HTML file from the command line as follows:

```
$ firefox phpmetrics_report/index.html
```

Alternatively, see the following:

```
$ chromium phpmetrics_report/index.html
```

Understanding the report

If you have opened a *PhpMetrics* report for the first time, you will find a wide variety of information. We will not dive into every little detail but will show you which parts of the report we think are the most valuable to start with.

To illustrate the usage of *PhpMetrics* better, we randomly chose an existing open source package called thephpleague/container as a code base to work on. It is an excellent PSR-11-compliant dependency injection container, which is just the right size to use as an example. *Figure 8.3* shows the overview page of an example report that we generated for it:

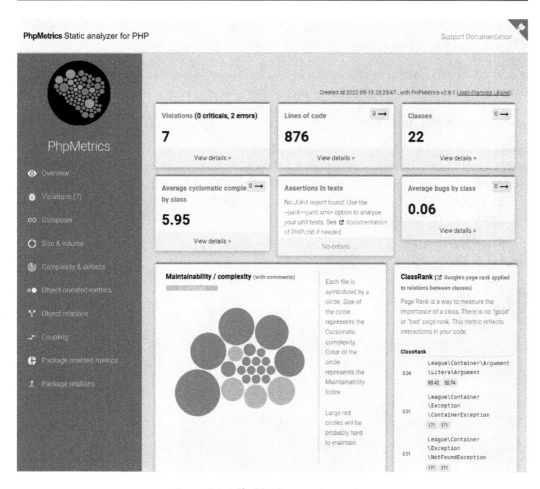

Figure 8.3: A PhpMetrics report overview

Key metrics

On the left-hand side, you will find the menu where you can access other pages of the report. The top part of the page is populated with a couple of key metrics, where the most interesting ones are:

- **Lines of code** tells you more about the size of this project. By clicking on the label, you will be sent to another page with a detailed list of all the classes and their related size metrics such as LOC.

- **Violations** shows you the number of violations that *PhpMetrics* has discovered. Again, by clicking on the label, you will be sent to another page with a list of classes and their violations – for example, if they are too complex (*Too complex method code*), have a high bug probability (*Probably bugged*), or use too many other classes or other dependencies (*Too dependent*).

- **Average cyclomatic complexity by class** tells you exactly what it says on the tin. The detailed view gives you more information about the complexity on a class level.

The other boxes offer interesting information as well, but the preceding ones are already perfect for getting a quick view of the most problematic parts of the code.

Maintainability or complexity

Underneath the key metrics, *PhpMetrics* shows a diagram, among other things, which you surely already spotted when first opening the report: the **Maintainability / complexity** graph. It consists of a colored circle for each namespace of the project, where the size of the circle represents the CC of a class. The bigger the circle, the higher the complexity. The color shows you the maintainability index, ranging from green (high) to red (low).

If you hover your mouse over a circle, you can see the namespace the circle represents and the two metrics in detail:

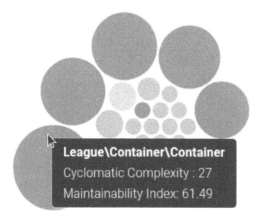

Figure 8.4: The Maintainability / complexity graph with a popup

This graph is extremely useful for quickly grasping the overall code quality – the fewer big red circles there are, the better. This way, you can see the problematic parts of your code easily.

Object relations

When you select **Object relations** from the left-hand menu, a graph that shows the relations between each namespace will appear. Hovering the mouse pointer over a text label will highlight its relations. The graph becomes massive, so we can not show it to you in this book in its full beauty, but we can at least give a first impression:

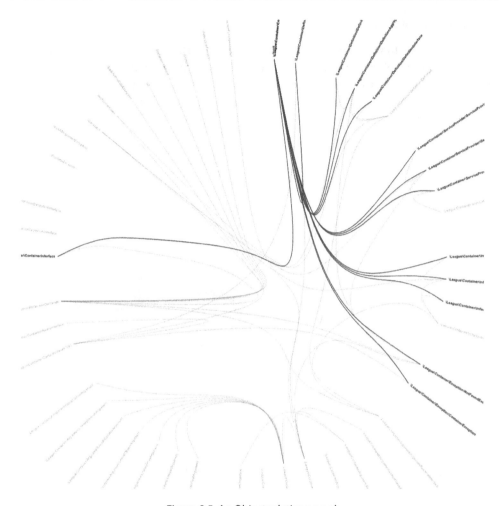

Figure 8.5: An Object relations graph

Coupling

The coupling of classes states how they depend on each other. There are two main metrics:

- **Afferent couplings** (**Ca**) tell you the number of classes that depend on this class. Too many dependencies indicate the importance of a class for the project.

- **Efferent couplings** (**Ce**) give you an idea of how many dependencies a class uses. The higher this is, the more the class depends on others.

Package-oriented metrics

The last graph we want to show you is the **Abstractness vs. Instability** graph. As the name already indicates, it shows the relationship between abstractness and the instability of packages. It was introduced by Robert Martin and is based on his work on object-oriented metrics. *Figure 8.6* shows you an example:

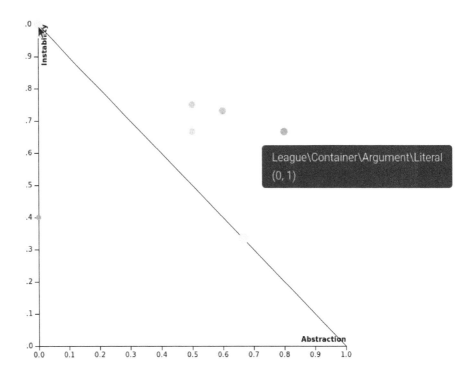

Figure 8.6: An Abstractness vs. Instability graph

But what exactly do these two terms mean in the context of software development? Let us look at the following definitions:

- **Abstractness (A)** is the ratio of abstract base classes and interfaces to the number of total classes in a namespace or package. The more these abstract types are included in the package, the easier and less risky the changes become. A ranges from 0 (concrete) to 1 (abstract).

- **Instability (I)** tells you how vulnerable a package is to change, expressed through a ratio of the Ce to the total Ce and Ca $(Ce + Ca)$. In other words, the more dependencies it has, the less stable it will be. I ranges from 0 (stable) to 1 (unstable).

Martin stated that packages that are stable and thus highly independent of other classes should also have a high level of A. Vice versa, unstable packages should consist of concrete classes. So, in theory, the A of a class weighs out its I. This means that ideally, A plus I should be 1 ($A + I = 1$). This equation also draws the angled line from the top-left to the bottom-right corner of the graph. You should strive to keep your packages close to the line.

In the actual report, you will find a table below the graph where you will find the values in more detail. If you hover the mouse pointer over a circle, a popup will appear that tells you the name of the class the circle represents, as well as the A (the first digit) and the I (the second digit).

Other information

This ends our tour through *PhpMetrics*. There is a lot more to discover, such as, for example, the *ClassRank*, where the famous *PageRank* algorithm from Google is used to rank the classes according to their importance (i.e., the number of interactions they have with other code parts). We can't cover everything in this book – however, by now, you already know many of the metrics. Its documentation is quite helpful to you. You will find a link to it on every page in the upper-right-hand corner.

The pros and cons of using metrics

In the two previous chapters of this book, you have learned about many tools and metrics that exist solely to help you write better software. The knowledge and wisdom and countless hours of endeavor on the part of hundreds if not thousands of software engineers can be added to your project in a matter of minutes.

The other side of the coin is that you might already feel completely overwhelmed by the sheer number of possibilities. Which tools should you choose? Which metrics should you focus on in the future?

If you have that feeling already, do not worry. We will not leave you alone in all this mess but help you find a setup that fits your needs during the next chapters. To begin with, let us take the time to look at the pros but also cons of using code quality metrics.

The pros

First, each software project is a unique piece of work. It grows based on certain circumstances, such as the skill sets of the developers and the available packages or frameworks at that time, but also external factors, such as deadlines, which often enough affect code quality negatively.

Code metrics help you to get an overview of the current state that a project is in. If you, for example, take over a project made by a former team member, you want to know what awaits you. By having an idea of the code quality, you can immediately adjust your estimated efforts on future tickets in whatever direction.

Code quality metrics also help you understand where code needs to be improved. It is excellent training to refactor your code, and by using the metrics, you know when you have succeeded. Regardless of whether you are working on your own pet project, you want to contribute to an open source project, or you work in a team, it is always a nice achievement to finally get some more green lights on the reports.

If you found a piece of code that urgently needed refactoring for a valid reason, but your project manager did not want you to do it, you could use the metrics to show them how terrible things are and that it is just a judgement based your own opinion. Code metrics are unbiased and (painfully) honest.

Finally, another important use case of these metrics is to prevent you from writing bad code in the first place. Sometimes, it might be a bit annoying to write code that adheres to all these rules but be assured that the effort pays off eventually.

The cons

Previously, we said that deadlines can harm code quality because they keep us from refactoring code smells or adding more tests. While this is true, we must be aware that once they start by measuring the quality of their code, some developers start to refactor a lot more code than necessary because they get rewarded with better metrics. Why is that problematic?

For example, imagine there is a class in your current project that has a low maintainability index, high NPath complexity, and just by looking at it, you can immediately see how bad it is. However, it has matured over time, being fixed so often that at some point, it has proven to work without bugs. Now, your tools tell you that this class is of bad quality. Should you still jump on it and start refactoring it?

There is, of course, no clear yes or no. As mentioned previously, if you work on code in your spare time, it makes sense (and is fun, too) to refactor a class to remove most of the code smells. If you are working on commercial projects, as in, working as a software engineer for a living, you will not always have the time to do so. There are bugs to squeeze, which make the users of your software unhappy, while on the other hand there are features to implement, for which they are desperately waiting. Overall, it is the satisfied customers who pay your bill. Finding the sweet spot between development speed and code quality is never easy – you just need to be aware that you sometimes have to take the bitter pill and leave bad code alone for now.

Do not use metrics to compete against colleagues, or worse, to talk badly about former developers who left you alone with the project. Be aware that everybody works as well as possible, based on their skills. Nobody deliberately tries to write bad code – often enough, it happens because the developers have never heard about clean coding principles or they were under such sheer time pressure that they had to do copy and paste coding to make their managers or customers happy. Your work environment should be a place of respect, helpfulness, and tolerance, not competition.

Summary

This chapter introduced you to some of the most used code quality metrics in the PHP world. Furthermore, we presented you with the tools that help you gather them. Of course, there are many more that we could not cover in this book, but you do not have to know them all – you are now equipped with a solid understanding of code quality metrics that will help you in your daily working routine.

Code quality tools and metrics are surely no silver bullet for all problems. On the one hand, they can be extremely helpful for improving your code. On the other hand, you should not take them as the ultimate measure. There are numerous examples of successful types of software that would never pass these quality checks, such as WordPress. Be sure, though, that the creators of WordPress would have done things differently if they had known beforehand.

In the next chapter, we will leave the realm of theory. We will learn how to organize the tools that we introduced in the last two chapters into our projects. Every project is unique, so we will offer you different flavors to fit your needs.

Further reading

- **dePHPend** (`https://dephpend.com/`) is a tool that can draw UML diagrams for your PHP code and be used to spot problems in your architecture.

9

Organizing PHP Quality Tools

In the last two chapters, you learned a lot about quality metrics and how to measure them. There will surely be a couple of tools you would like to use in your future work environment, and these tools work best if they are seamlessly integrated so that you don't even have to think about using them anymore.

Therefore, in this chapter, we will show you how you can organize these tools in a way that they will be the most productive and helpful in your daily work. This includes the following topics:

- Installing code quality tools using **Composer**
- Installing code quality tools as `phar` files
- Managing `phar` files using the **PHAR Installation and Verification Environment** (**Phive**)

Technical requirements

If you followed the examples in the previous two chapters, you do not need to install anything else. If not, please go back to those chapters and install all the necessary tools first.

All code samples can be found in our GitHub repository: `https://github.com/PacktPublishing/Clean-Code-in-PHP`.

Installing code quality tools using Composer

Most **PHP: Hypertext Preprocessor** (**PHP**) projects nowadays use Composer for a good reason. Before it entered the PHP world in 2012, keeping all external dependencies (that is, code from other developers) up to date required a lot of manual work. The required files had to be downloaded from the corresponding websites and added to the correct folders of the project. Autoloading (that is, the automatic resolving of file paths from the class name) was not standardized, if available at all. So, usually, the wanted classes needed to be actively imported using `require()` or `require_once()`. If there were any conflicts between package versions, you had to somehow solve the issues yourself.

Composer greatly simplified these efforts by solving these issues. It introduced a central repository called **Packagist** (`https://packagist.org`), where all available packages are hosted. Furthermore, it fixed the version problem by introducing version constraints. These are rules that tell Composer which versions of other packages a package supports, enabling Composer to often automatically resolve the right versions. Another groundbreaking feature was the support of autoloading for installed packages. There, we now usually use `require()` only to import Composer's autoloader.

All these features helped PHP compete with other web languages such as Python or Ruby, and without it, PHP would probably not be the most widely used language on the **World Wide Web** (**WWW**) anymore, as it is today. Therefore, we want to give Composer the space in this book that it deserves. In this section, we will show you the most used installation method. Additionally, we will also have a look at another, lesser-known way of using Composer in your project.

Installing code quality tools using require-dev

Throughout the last chapters, we already used Composer to install tools most of the time, so by now, you should already be familiar with the most common use case there is: adding dependencies to your project. **Dependencies** are code packages written by other developers that can quickly be integrated into your project.

To recap, this is done using the `require` keyword and the package name. For example, if you want to add PhpMetrics, you can do so by running the following command:

```
$ composer require phpmetrics/phpmetrics --dev
```

Typically, packages are identified by the name of the developer(s)—the so-called *vendor*—and, separated by a slash, the package name. In the preceding example, the vendor and the package name are identical, but this is not always the case.

Let us look at the `--dev` option in more detail now. When we run the `composer require` command with this option, Composer will add the package in another section of the `composer.json` file, called `require-dev`. Here, you can see an excerpt of a typical `composer.json` file:

```
{
  "name": "vendor/package",
  ...
  "require": {
    "doctrine/dbal": "^2.10",
    "monolog/monolog": "^2.2",
    ...
  },
  "require-dev": {
```

```
    "phpunit/phpunit": "^9.5",
    "phpmetrics/phpmetrics": "^2.8",
    ...
  },
  ...
}
```

The idea behind the `require-dev` section is that all packages within this section are not necessary to run the application in production. In the local environment, or during the build, you will surely need PHPUnit and all our held-dear code quality tools; in production, they are not needed anymore, though.

In fact, you should strive to have as few packages as possible in production. This is mainly for two reasons, as follows:

1. Each package you add will be included in Composer's autoload mechanism, which costs performance on every request. Internally, Composer builds a so-called *classmap*, which is a simple array that maps a class name to the respective file location. If you are curious about this, check out—for example—the `vendor/composer/autoload_classmap.php` file. Depending on the number of packages your project uses, this file can get huge, slowing down your application.

2. Every additional package can possibly introduce security issues. The less code there is, the fewer attack vectors there are.

By default, Composer will install all dependencies. Thus, be sure to run it using the `--no-dev` option for your production builds to exclude the packages in `require-dev` from being installed. In your local environment, however, you do not need to worry about anything else at this point.

The previously described installation method is the one you will encounter the most for good reasons: it does not require any additional tooling and there's just one additional option to be used when installing it on production. This makes it a perfect starting point and is often fully sufficient for a small project. Another approach worth knowing is the global installation of Composer, which we will discuss in the next section.

Global installation

If you are working on numerous projects simultaneously on your local system, you can choose to install Composer and packages globally, which means that they will not be installed in any project `root` folder and will thus not be added to any `composer.json` file there. Instead, both Composer and the packages will be installed in a single folder, which is usually `~/.composer`. In this folder, you will then find another `composer.json` file that keeps track of the globally installed packages, as well as another `vendor` folder, where their code is installed.

Installing packages globally simply requires adding the `global` modifier, like so:

```
$ composer global require phpmetrics/phpmetrics
```

Likewise, updating all global packages is effortless too, as demonstrated here:

```
$ composer global update
```

After global installation, tools such as the **PHP Coding Standards Fixer** (**PHP-CS-Fixer**) can simply be executed without having to specify the path, like so:

```
$ php-cs-fixer fix src
```

To make this approach work, however, you will need to add this global folder to the execution path. Please see the Composer documentation (`https://getcomposer.org/`) for more details on how to do this for the operating system you use.

Using the global installation feature should only be chosen if you are working alone on your projects and do not use any build pipelines. If you are working in a team and/or make use of a **continuous integration** (**CI**) pipeline, you should install it for every project separately, though.

According to common best practices such as the **Twelve-Factor App** principles (`https://12factor.net`), all dependencies should be explicitly declared, and no global dependencies should be relied on since you can never be sure which version will be installed. Although code quality tool packages are not part of the actual program code, they are still part of the build process. Even small differences between the installed versions can lead to unforeseen behavior and can generate confusion when errors cannot be reproduced locally.

Furthermore, you want to make the initial installation of a project as easy as possible. Having your teammates install all the required tools manually is a time-wasting and error-prone process that can lead to frustration.

For the aforementioned reasons, we do not encourage using the global installation method.

Composer scripts

Once you have decided on a possible way to install Composer and have used it to download your desired tools, you want to start using them in the most straightforward way possible. In *Chapter 11*, *Continuous Integration*, where we will talk about CI, we will also show you how to run these tools automatically during the build process. For now, however, we want to show you how Composer can assist you in running them manually whenever required.

Let us consider the following example: as a first step, we would like to run PHP-CS-Fixer to automatically fix the code in our `src` folder. Afterward, we wish to run PHPStan with level 1 on our code as well. You could surely run both steps separately, but we want to add a bit more comfort and execute both tools in one go.

To achieve this, we can utilize the `scripts` section of the `composer.json` file *in the project root*. There, we have to add the tools we want to execute under a concise command name, such as `analyze`. The following example shows what this could look like:

```
{
    ...
    "scripts": {
      "analyze": [
        "vendor/bin/php-cs-fixer fix src",
        "vendor/bin/phpstan analyse --level 1 src"
      ]
    }
}
```

We used the **JavaScript Object Notation** (**JSON**) array notation here to add each command in a separate line, which makes it easier to read and maintain than writing everything in one line.

If you want to share these Composer commands, you might want to add a short description text as well, which is displayed when you execute `composer list` to see a list of available commands. To do that, you need to add the `script-descriptions` section to your `composer.json` file. For the previously introduced `analyze` command, it could look like this:

```
{
    ...
    "scripts": {
        ...
    },
    "scripts-descriptions": {
        "analyze": "Perform code cleanup and analysis"
    }
}
```

By installing the tools in a subfolder, we found a suitable way to organize our code quality tools without letting them interfere with our application dependencies. But what if, for whatever reason, you are not using Composer in your project, or you dislike the fact of having two `composer.json` files in your repository? In the next section, we will introduce an alternative way that does not make use of Composer.

Installing code quality tools as phar files

Composer is not the only possible way to add code quality tools to your project. In this section, we will show you to add the tools as `phar` files.

We already came across **phar** in the previous chapters. It stands for **PHP Archive** and can be considered a whole PHP project within a single file. Technically, it is a self-executable ZIP archive that contains the source code of the application plus all necessary dependencies as well. The big advantage is that the required code is available immediately after download, so you can instantly use any `phar` file right away without having to care about Composer or dependencies at all. Furthermore, `phar` files are supported by all modern PHP versions.

This makes the usage of `phar` files quite handy, as you can treat them like binaries. Usually, you can download the many PHP tools we introduced to you so far as `phar` files directly and place them in whatever directory you want. However, there is no unified way these files are provided, so please refer to the official documentation of each tool.

Let us have a look at how to do that for the `phploc` tool, which we introduced in *Chapter 7, Code Quality Tools*. According to its GitHub repository, you can simply download it from the *PHPUnit* website, as they are both from the same author. The following code snippet shows how you can do this:

```
$ wget https://phar.phpunit.de/phploc.phar -O phploc
```

Note that we install the tool under the name `phploc`, and not `phploc.phar`. The `-O` option allows you to specify a different filename than the one you are downloading. The `.phar` extension is not necessary to execute the tool, so you can save on a bit of typing effort here.

> **Phar and checksums**
>
> Downloading and executing files from the internet always includes the risk that they can be corrupted and infected with malicious code. That is why the authors of the tools often generate checksums (for example, through hash algorithms such as **Secure Hash Algorithm 256 (SHA256)**) of the downloads and publish them on their websites so that you can use them to verify the integrity of the download. Please check the official websites of the tools you're considering using to find out if they offer checksums, and how to verify them.

Of course, you can download it using any method you like, be it `curl` or via the browser. Once you download it, you can immediately run it using your local PHP installation, like so:

```
$ php phploc src
```

If you do not want to type `php` every time, you need to make the `phar` file executable, which on Linux—for example—would look like this:

```
$ chmod +x phploc
```

Afterward, you just need to run the following command to execute `phploc`:

```
$ ./phploc src
```

Keeping your phar files organized

Now, we do not just want to download the `phar` files—we also want to keep them organized in our project so that any other developer does not have to do any manual work before using them. The most obvious choice is to add these files to your repository, and that is precisely what we will look at now. In the following example, we will use Git, but this approach would work with any other **version control system** (**VCS**).

It is generally discouraged to store large files in Git because they can affect performance negatively. GitHub, for example, blocks files that are greater than 100 **megabytes** (**MB**). However, the `phar` files we use are usually just a few MB in size, so adding them should not have any negative side effects.

> **Git Large File Storage (Git LFS)**
> If you need to store large files in Git, consider using Git LFS, which was designed exactly for this use case. For our needs, though, we do not have to use it.

You can freely choose where to add `phar` files to your project. A common place is the `root` folder; however, since this will get quite crowded over time, we recommend using a separate folder to store them. A good place would be the `tools` folder again, just like we used it in the previous section. You do not need to consider anything else; just add them to the repository like any other file.

Let us assume you copied the `phploc` file into the `tools` folder and made it executable as described previously. Then, you would just execute as follows:

```
$ tools/phploc src
```

Using `phar` files is easy and does not interfere with your application dependencies. However, they are not perfect: if you want to update them, you need to look up the download **Uniform Resource Locator** (**URL**), download the `phar` file, and validate its checksum manually every time—for each tool. In the next section, we will show you how to ease that process by introducing another dependency management tool: Phive.

Managing phar files using Phive

In the previous section, we learned about using `phar` files instead of using Composer to install our code quality tools. This approach works fine, but it does require some extra work in case you want to update them.

Phive is a tool that takes over that extra work. Let us install it right away.

Naturally, Phive itself can be downloaded as `phar`, too. The following commands will download it under the name `phive` and make it executable:

```
$ wget https://github.com/phar-io/phive/releases/
download/0.15.1/phive-0.15.1.phar -O phive
$ chmod +x phive
```

Please note that this installation method is not very secure. Check the tool's website (`https://phar.io`) to learn how to install it securely and how to make it globally available.

For demonstration purposes, the simple download works just fine. Once the file is downloaded and made executable, you can directly start using Phive to install the first tools. Let us use `phploc`, which we introduced in the previous chapter, to demonstrate how it works, as follows:

```
$ ./phive install phploc
```

> **Download verification**
>
> Phive not only takes care of the *installation* but also of the *verification* of the downloads. This is done automatically during the installation process. However, this requires the vendor to make the checksums available, which is also the main reason why not all tools can be managed through Phive.

As you saw previously, installing a tool is done just by using the `install` command. The following four things have happened now:

1. Phive downloaded the latest version of `phploc` and verified its checksum.

2. The `phar` file got stored in a shared folder (usually located in your home folder under the name `.phive`).

3. Phive then created a symbolic link to that shared folder. A symbolic link is a reference in the filesystem so that a file or directory can appear in multiple directories, although it is stored in just one place. By default, this symbolic link is stored in the `tools` folder, which will be generated if it does not exist.

4. Another `.phive` folder has been created in your project root folder, which is used to store the information about which tools have been downloaded.

Symbolic links appear just as "real" executables in your directory, while the original file stays in one single location. If you do not want to use symbolic links, you can install a file copy instead by using the `--copy` option.

After installation, the execution of `phploc` is simple, as we can see here:

```
$ tools/phploc src
```

Phive offers more useful commands. Just run the following code without any command to get a list of them:

```
$ ./phive
```

Here, we introduce the most important ones:

- `list`—Lists all tools that can be managed through Phive
- `update`—Updates all installed `phar` files, if newer versions are available
- `selfupdate`—Updates the `phive` executable itself
- `outdated`—Tells you which `phar` files can be updated
- `status`—Lists an overview of all installed tools

Adding Phive to your project

If you work in a team, you not only want to locally install the `phar` files. Phive has got you covered here as well. Two steps are necessary to properly add Phive to your project, as follows:

1. Add the `.phive` folder in your project root to your repository. The `phars.xml` file inside contains all necessary information (such as the `composer.lock` file).
2. Make sure the `tools` folder is not under version control (for example, by using a `.gitignore` file). You explicitly do not want to add the `phar` files themselves to your repository.

Once this is done, the next time the project gets checked out from the repository, the tools can be installed by executing the following command:

```
$ ./phive install
```

This command can be easily integrated into other workflows—for example, as an additional `post-install-cmd` script in the `composer.json` file.

That is already all you need to know about Phive to start using it. As usual, we recommend you read the official documentation because we could not cover every feature it provides in this book.

Summary

Composer is an indispensable part of today's PHP world. The usual approach to adding code quality tools to your project is by adding them to the `require-dev` section of the dependencies, which works fine in many cases.

However, Composer is not the one and only way there is. Therefore, in this chapter, we introduced two more options to manage your code quality tools: by adding the `phar` files manually to your project, or by utilizing Phive to manage the `phar` files.

You are probably eager to apply all your gained knowledge to your code now. However, relentless refactoring can do more harm than good, and clicking through all parts of your application after every change to check if anything broke will cost you a lot of time and can be very frustrating. Thus, in the next chapter, we will show you how automated testing can help you here.

10

Automated Testing

If you've read all the chapters of this book from the very beginning, you'll have not only an idea of the theoretical background but also a good set of tools at hand that will help you write great **PHP: Hypertext Preprocessor (PHP)** code. Of course, you can just go and refactor all the code there is, probably using some automated code manipulation capabilities our tools offer.

You will not write perfect code on the first try—it usually takes several iterations until you are satisfied. And since you never stop learning, you will refactor parts of your code even months or years later. Yet even the most sophisticated code quality tools will not prevent you from having to do one tedious task: testing your code after you've made changes to ensure that it still works as expected. That is why in this chapter, we want to introduce you to **automated testing**.

Through automated testing, you will be able to verify that your improvements to the code did not break its functionality in a fast and reliable way. This is one of the cornerstones of writing clean code since it enables you to refactor code with confidence.

The topic of automated testing deserves a whole book or two, so we can only scratch the surface here. Yet since we are convinced you will greatly benefit from it in your daily work, we hope that this chapter will make you want to learn more about this exciting topic.

The following sections will give you a good overview:

- Why you need automated tests
- Types of automated tests
- About code coverage

Technical requirements

Additional to the technical requirements of the previous chapters, you will need to install the **Xdebug** PHP extension. We will provide you with more information on that topic in the corresponding section, *About code coverage*, later in this chapter.

The code files for this chapter can be found here: `https://github.com/PacktPublishing/Clean-Code-in-PHP`

Why you need automated tests

Although **PHPUnit**, the standard unit testing framework for PHP, has existed since 2006, automated tests are still not used in all PHP projects today. A lot of potential is wasted here because automated tests have many benefits, such as the following:

- **Speed and reliability**: Imagine you need to execute the same testing steps over and over. Soon enough, you would make mistakes, or just skip the tests at some point. Automated tests, however, do the boring work for you in a much faster and more reliable way—and they do not complain.

- **Documentation**: With automated tests, you can indirectly document the functionality of code through assertions, which explain what the code is expected to do. Compared to comments or articles in a wiki, you will immediately be notified by the failing tests when something has changed significantly. We will discuss this topic again in *Chapter 13, Creating Effective Documentation,* when we talk about creating effective documentation.

- **Onboarding**: A test suite with good coverage of our code will help new developers to get productive on a project faster. Not only do the tests act as additional documentation, but they also let the developers make changes or add features with confidence. They can verify that their changes do not break anything before they get deployed to any staging or production environment.

- **Continuous integration/continuous deployment (CI/CD)**: Be it CI or CD, if your tests are automated, you can trust through your build pipeline that the code you merge is not broken, which enables you to push code to production faster, and thus more often. In the next chapter, we will have an in-depth look at this topic.

- **Better code**: You do not strictly have to follow the infamous **test-driven development** (TDD) approach to benefit from tests already in development. Writing unit-testable code even improves your code. To be able to test code in isolation (for example, without having a real database running in the background), you need to write your code with separation in mind. If external dependencies are injected using **dependency injection** (DI), they are much easier to replace with test objects than if you instantiate them in the class functions. We will have a closer look at the **DI pattern** in *Chapter 12, Working in a Team*. Additionally, long and complex functions are equally hard to test as short ones (think of—for example—the **NPath complexity** here, which we discussed in *Chapter 8*), so you will very soon start to write shorter functions to reduce the number of decision paths in your code.

- **Easier refactoring**: Automated tests are an invaluable tool when you want to refactor a project based on the results of the static code analyzers we introduced back in *Chapter 7*. You can apply their recommendations or even the automated code fixes, and just after running the tests, you will know if this introduced any side effects or not. Since refactoring is the most important use case for us in the context of this book, we will discuss it in more detail in the next section.

> **TDD**
>
> TDD is a way of developing code that combines writing tests and the actual code at the same time. The basic idea is simple and often referred to as **red/green/refactor**: before you write any code for a new feature or even a bug fix, the first step is to write a test that checks the expected outcome. Since you have no actual code written yet, the tests will fail (indicated by the color red). In the second step, you write the code, without paying too much attention to making it perfect, until the tests pass (green). Since you now already have working tests, you can easily improve (refactor) the code until you are satisfied.
>
> The TDD paradigm ensures that you will have all your code covered with tests and that the code is already written in a fully testable way. Do not take things too seriously, though: there are times when you just want to experiment without having a clear goal in mind—for example when you play around with a new **application programming interface (API)**. In this case, you do not need to follow TDD.

Easier refactoring with tests

If you start a project *on the green* (that is, you write it from scratch), you can have the comfort of getting immediate feedback from your code quality tools as soon as you start coding it. This is a great help, yet even the best tools will not prevent you from making bad decisions and writing code that you will want to undo at some point.

That happens to everyone and should not discourage you at all. You learn something new every day, and while your personal skills evolve, so will your code. If you look at your code from 1 year ago, you'll probably want to refactor it right away.

And of course, not only your skills, but the whole PHP ecosystem constantly improves. Many things that are standard today were simply non-existent back in the day. New packages or language features are constantly being introduced, and you want to use them in your projects to not stick with the old techniques forever.

So, code changes over time—that is completely normal, and we as developers should embrace change; no piece of our code will ever be final. We refer to changing existing code as **refactoring**. The interesting part about refactoring is that code gets changed, but the software appears unchanged to the user. All work happens "under the hood". If you—for example—updated the framework of your project to the latest version, and the users noticed no direct changes, then you did your job well.

Refactoring has benefits; otherwise, we would not do it. If done right, it can lead—for example—to improved performance, increased security, or generally allows an application to be scalable in the cloud. And yet, refactoring often comes with a bad connotation. Managers in particular tend to think that refactoring means changing code just because there is yet another hype in the web development world the engineers want to follow, and precious working hours are wasted.

Let us be honest: of course, this happens too. The line is often difficult to draw. For example, imagine your duty is to maintain an old but perfectly working PHP application that uses the **Singleton pattern** for object instantiation. If you only have to do small changes occasionally, there actually is no need to refactor it to use DI. However, if you are required to implement ongoing changes, such as adding new modules and tests for it, it might be a good choice to do so.

Often, you will have to justify your refactoring work. It is then helpful to rather speak of **system health maintenance** on the code instead. Everybody is completely fine with the fact that machines require maintenance: parts get replaced, the lubricant must be renewed, and so on. Yet, for some reason, our software ought to work forever.

Having good arguments for refactoring now, we want to understand how testing can help us here. To accomplish that, let us have a closer look at the different test types there are in the next section.

Types of automated tests

Although **unit tests** are probably the most known type of automated testing, there is more to discover. In this section, we will introduce the most common (and important) test types. A well-known testing concept is the **testing pyramid**, which is shown here:

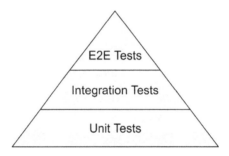

Figure 10.1: Testing pyramid

This concept basically shows three types of tests—namely, **end-to-end tests** (or **E2E tests** in short), **integration tests**, and **unit tests**. We will explain each test type and its position in the testing pyramid in the following sections.

Unit tests

As the name implies, **unit tests** are about testing small units of code. It is best practice to write one test for each functionality of an object only; otherwise, the tests will become bigger and harder to understand and maintain. This keeps the tests small as well, and that is why there is usually a lot of them. Depending on the project size, having hundreds or thousands of unit tests is completely normal, so it is important to keep them executing as fast as possible. Usually, they should not take longer than a few microseconds each.

Unit tests should run in isolation, which means that in the tests, the tested objects do not interact with any other external services, such as databases or APIs. This is achieved by faking the external dependencies, which is called **mocking** in unit testing jargon. Simply speaking, we replace external objects—such as services or repositories that are used within our test object—with **mock objects** (or **mocks**, in short). These objects simulate the behavior of the dependencies they replace during the runtime of the unit tests. This ensures that a test does not suddenly fail just because—for example— some data in the database, which our test was relying on, has changed.

Because tests of this type are small, fast, and do not rely on external dependencies, it is relatively easy to create a test setup for them. They are extremely helpful because they can tell you within seconds if your last changes to the code caused any problems or not. That is why they are the foundation of the testing pyramid.

If you are new to testing, it makes sense to start with **PHPUnit**, as it is the industry standard in the PHP world. If you start on a new project, it is most likely that PHPUnit will be used. There are other testing frameworks that have their unique advantages, such as **Pest** (https://pestphp.com). Once you have grasped the concept of unit testing with PHPUnit, we encourage you to give them a try as well.

A drawback of unit tests is the fact that they do not interact with each other. This may even lead to having all your tests passing while your application is broken, just because the interaction between the classes was not tested properly.

To illustrate this problem, we create an elementary demo application. Let us have a look at the most important parts of it.

Demo application source code

You will find the full source code in the GitHub repository to this book:

`https://github.com/PacktPublishing/Clean-Code-in-PHP`

First, we create a rudimentary class, called MyApp, as follows:

```php
<?php

class MyApp
{
    public function __construct(
        private myRepository $myRepository
    ) {
    }

    public function run(): string
```

```php
    {
        $dataArray = $this->myRepository->getData();

        return $dataArray['value_1'] .
          $dataArray['value_2'];
    }
}
```

The MyRepository method gets injected through the constructor. The only method, run, uses the repository to fetch data and concatenate it. It is important to note that MyClass expects a certain array structure to be returned by MyRepository. This is not recommended to do, but you will still find this a lot "in the wild". Therefore, it serves perfectly as a demonstration.

MyRepository looks like this:

```php
<?php

class MyRepository
{
    public function getData(): array
    {
        return [
            'value_1' => 'some data...',
            'value_2' => 'and some more data'
        ];
    }
}
```

In real life, MyRepository would fetch the data from an external data source, such as a database. For our example, it returns a hardcoded array. If the run method of MyClass gets executed, it will return a some data...and some more data string.

Of course, we also added tests (using PHPUnit) for the preceding classes. For brevity, we will only show the test cases in the following code snippet, not the whole test classes:

```
public function testRun(): void
{
    // Arrange
    $repositoryMock =
        $this->createMock(MyRepository::class);
    $repositoryMock
        ->expects($this->once())
        ->method('getData')
        ->willReturn([
            'value_1' => 'a',
            'value_2' => 'b'
        ]);

    // Act
    $appTest = new MyApp($repositoryMock);
    $result = $appTest->run();

    // Assert
    $this->assertEquals('ab', $result);
}

public function testGetDataReturnsAnArray(): void
{
    // Arrange
    $repositoryTest = new MyRepository();

    // Act
    $result = $repositoryTest->getData();

    // Assert
    $this->assertIsArray($result);
    $this->assertCount(2, $result);
}
```

> **The Arrange-Act-Assert (AAA) pattern**
>
> You may have noticed that we added three lines of comments to both of our test cases: `Arrange`, `Act`, and `Assert`. We did that to demonstrate the most used pattern to write unit tests: the **AAA pattern**. Even without having ever written a single unit test yourself, it helps you to understand how they work.
>
> Firstly, the test objects and required prerequisites such as mock objects get prepared (`Arrange`). Secondly, the actual work of the object under test is executed (`Act`). Finally, we ensure that the results of the test meet our expectations (`Assert`). If one of the assertions is not met, the whole test fails.

Two things are noteworthy here, as set out next:

1. In `testRun()`, we create a `$repositoryMock` mock instead of using the actual `MyRepository` method. This is because we assume that `MyRepository` would normally fetch the data from an external data source, and we do not want to write unit tests that have external dependencies.

2. `testGetDataReturnsAnArray()` does not test the repository very well. We just check if the result is an array and that it has two entries, but not which array keys are returned.

Now, imagine that for whatever reason, one fellow developer decided that the `value_1` and `value_2` array keys were too long, and renamed them `val1` and `val2`. If we now execute our application, it will of course break, as illustrated here:

```
$ php index.php
PHP Warning:  Undefined array key "value_1" in
  /home/curtis/clean-
  code/chapter10/unit_tests_fail/src/MyApp.php on line 18
PHP Warning:  Undefined array key "value_2" in
  /home/curtis/clean-
  code/chapter10/unit_tests_fail/src/MyApp.php on line 18
```

However, if you execute the tests, they will still pass, as we see here:

```
$ vendor/bin/phpunit tests
PHPUnit 9.5.20 #StandWithUkraine


..                                                        2 / 2 (100%)


Time: 00:00.008, Memory: 6.00 MB
```

```
OK (2 tests, 4 assertions)
```

This illustrates that having unit tests is important, but it does not necessarily mean that we will not introduce bugs anymore because they can be faulty or they test the wrong things, as in our example.

Often enough, objects such as **repositories** that interact with external systems are not tested at all because it requires a more complex test setup—for example, using an additional test database with fake data. If we just replace such an object with a mock, the test will work correctly. If there are significant changes on the original object later and the mock does not get updated to reflect those changes, we can end up in a situation like we just described.

To overcome this problem, we need a way to additionally test our classes without replacing dependencies with mocks. For that, we will introduce the second test type of the testing pyramid—integration tests—in the next section.

Integration tests

The second types of test we want to look at are **integration tests**. Unlike unit tests, which are supposed to not use any external dependencies, with this test type we want to do the opposite: we want to test code as it would normally run, without replacing anything with mocks.

You might have already witnessed unit test suites that made use of a test database or some external API. Technically speaking, these tests are not unit tests anymore, but integration tests (or **functional tests**, as they are also called). We could theoretically use PHPUnit for these tests as well, or use particular testing tools that take over a lot of groundwork for us.

The following code snippet shows an example of an integration test:

```php
public function productIsSaved(Tester $tester)
{
    $product = new Product();
    $product->setId(123);
    $product->setName('USB Coffee Maker');
    $product->save();

    $this->tester->seeInDatabase(
        'products',
        ['id' => 123, 'name' => 'USB Coffee Maker' ]
    );
}
```

The preceding function shows what an integration test could look like if we used the **Codeception** test tool (see the next information box for more information). An object called `$tester` gets passed into the test that is a `Helper` object that offers the functionality we need to perform—for example, database checks. After executing the `save` method on the `$product` test object, we use this `Helper` object to validate if the data we would expect has actually been written into the database.

> Codeception
>
> Codeception (`https://codeception.com`) combines a variety of test types, such as unit, integration, and even E2E tests in one tool. Under the hood, it is based on existing tools such as PHPUnit. It offers modules for all major frameworks and thus integrates well into most PHP projects.

Using integration tests makes the test setup more complex because we must make sure that the external dependencies we use are always in a reliable state. For example, if you need to rely on a certain user in your database, you must make sure that it always has the same data, such as the user **identifier** (**ID**) you test for; otherwise, your tests will fail. This usually requires creating a fresh test database before every test run, to make sure no leftovers from previous test runs disturb our test. Furthermore, we need to run **database migrations** to ensure that the test database schema is up to date. Finally, we have to fill it with test data, which is called **seeding**.

The main drawback of this test type is the execution speed. Database transactions are slow (compared to using mock objects), and we need to prepare the test database upon every test run. Integration tests also tend to break easier, or they become flaky (unstable) because the interaction with other dependencies quickly becomes very complex: if the previous test changed the database in a way the next test did not expect, your test run fails, although the code has not changed.

For example, a new test is added to the test suite that checks if a class updates a dataset in a certain way. Since this is an integration test, it will use the test database for it and change that particular dataset. After the execution of this test, however, the changed data will still be in the test database. If another test that runs *after* this new test is relying on the previous data, it will fail. Despite the added complexity of your test setup, integration tests ensure that the integration of tested objects within the application context works fine. That is why they should be an integral part of your test strategy, being the second layer in the testing pyramid.

You will find a lot of integration tests when it comes to testing repositories, models, or controllers. However, they cannot test the interaction between PHP and the browser. Since we use PHP mainly to build web applications, this is an aspect that we should not forget about. Luckily, the last test type from the testing pyramid covers exactly this problem.

E2E tests

For this test type, we will leave the realm of PHP for a moment. With **E2E tests**, we want to ensure that the whole flow from the server to the client (for example, the browser) and back again to the server is working alright. We basically simulate a user sitting in front of the computer and clicking through our application.

To achieve this, we first need a reproducible testing environment. Just as for integration tests, we must ensure that the application we want to test is always in the same state. This means that we need to ensure that the same set of data (for example, blog posts, or articles in a shop) is available on every test run.

Secondly, we need to automate the user interaction with our testing environment. Here is where things get interesting: we not only need the application, but also a local web server and a browser to run it in and simulate the user interaction. The web server adds more complexity to the test setup but is usually not a blocker. For user interaction, we need to use a so-called **headless browser**. Such a browser can interact with a server without having to open a browser window. This is an extremely useful feature because we can use it on the command line without having to install a full-fledged operating system with a **graphical user interface** (**GUI**) such as Ubuntu Desktop or Windows. This saves us a lot of installation time and helps us not to increase the complexity even further.

At the time of writing, **Google Chrome** is the preferred choice, as it is not only the most widely used browser engine nowadays, but also provides a **Headless** mode, or in other words, it can act like a headless browser. With modern frameworks such as **Cypress**, it is effortless to automate user interaction with our application. Think of it as a script that tells the browser which **Uniform Resource Locator** (**URL**) to open, which button to click, and so on. The following example shows a simplified Cypress test:

```
describe('Application Login', function () {
    it('successfully logs in', function () {
        cy.visit('http://localhost:8000/login')

        cy.get('#username')
            .type('test@test.com')

        cy.get('#password')
            .type('supersecret')

        cy.get('#submit')
            .click()

        cy.url()
            .should('contain',
```

```
                    'http://localhost:8000/home')
        })
    })
```

> **Cypress**
>
> The Cypress testing framework (`https://www.cypress.io/`) makes writing E2E tests very easy, as it handles the setup of and communication with the headless browser for you. Yes—the tests are to be written in JavaScript, but that should not stop you from giving it a try.

The `cy` object represents the tester, which executes certain steps. In the preceding code example, it firstly opens the login page of a fictive application, fills out the `#username` and `#password` fields in a login form, and submits it by clicking on the **#submit** button. As the last step, it checks whether the login was successful and if the tester has been forwarded to the home screen. All these actions are then executed in a real browser, running in the background. Using this technology, it is possible to write test suites that literally click through our application, just as a human being would. They do not just test the PHP code but also the frontend code—a JavaScript error, for example, will quickly break the tests. Even if you cannot fix the bug yourself, you can still inform the frontend engineers in your team that there is a problem.

The modern frameworks make it much easier to write tests than was the case with older technologies, such as Selenium. In fact, it is so comfortable nowadays that also people who are not developers but have a solid technical foundation, such as **quality assurance** (**QA**) engineers, can easily write their own test suites. This approach takes a lot of pressure from the teams because the developers have to write fewer tests, and the QA people can set up their tests as they need them without having to wait for the developers.

Of course, E2E tests have some drawbacks, which is why they are just the third layer of tests in the testing pyramid: the test environment is more complex and needs more work to set up, especially if a database or any external API is used. This test type is also the slowest since it involves the browser in addition to the test setups of the previous test types. Lastly, these tests can easily break because usually, the testing frameworks use **Hypertext Markup Language** (**HTML**) `id` and `class` attributes, or even **Document Object Model** (**DOM**) selectors, to navigate through the DOM and find elements to interact with. Thus, a small change on the DOM can quickly break your whole test suite.

> **Page objects**
>
> If you are interested in creating maintainable E2E tests, you should check out the concept of **page objects** (`https://www.martinfowler.com/bliki/PageObject.html`).

The testing pyramid in practice

With unit, integration, and E2E tests, you now know the three most important test types, and their pros and cons. The suggested approach of having unit tests as a huge foundation, a fair amount of integration tests, and—finally—some E2E tests is a good starting point.

However, you do not have to strictly follow it all the time, since every project is different: if you, for example, want to start testing a yet completely untested application, introducing unit tests would require a lot of refactoring effort to make the classes testable. This refactoring will most likely introduce more bugs in the beginning than it will solve.

In this case, starting with a good E2E test coverage will be quicker and safer. Once the main parts of the application can be automatically tested, you can safely start with the refactoring and introduce unit and/or integration tests. If your application breaks because of the necessary refactoring, your E2E tests have you covered.

At the end of this chapter, we will list some more test types for you to evaluate if you are interested. For now, the three test types we covered in this chapter are the most important ones and should be enough for you to get started.

There is one important question that we have not really covered yet, though: how much of your code do you really need to test? We will discuss this in the next section.

About code coverage

Now we've explored the different test types, you probably want to start writing tests right away. But before you put this book away to get coding, let us finish this chapter with the question of how much of your code you should test.

A part of the answer lies within the concept of code coverage, which we already briefly mentioned in *Chapter 8, Code Quality Metrics* when we talked about code quality metrics. Let us have a closer look at it now.

Understanding code coverage

Code coverage measures the proportion of code that is covered with tests. The higher the code coverage, the better—if there are more tests, the software is less likely to contain bugs, and it will be harder to introduce new ones unnoticed. Higher code coverage is also a possible indicator of better code quality—as we discussed in a previous section of this chapter, tested code must be written in a certain way, which usually leads to better quality.

Generally, the degree of coverage is expressed simply using the percentage of tested code—that is, from 0% (fully untested) to 100% (complete code coverage). But how can we measure code coverage? For that, we will use PHPUnit, as it can create a code coverage report for us. However, it requires an additional PHP extension for the code coverage functionality. For this chapter, we decided to use **Xdebug**, the standard PHP **debugger** and **profiler**.

Setting up Xdebug

Xdebug is an extension for PHP, so it requires to be loaded as a module. Since its installation is a bit more complex and mainly depends on the operating system you are running PHP on, please refer to the official documentation at `https://xdebug.org` on how to install and configure it. There are also plenty of tutorials available on the internet.

If you refactor your code, you probably want to know what performance implications the changes had. Did you improve the execution time, or did it get even worse? With a so-called **profiler**, you can measure the execution time of each function in detail and see where bottlenecks are hidden.

We cannot cover this topic in our book, but since over the course of this chapter we already worked with Xdebug, you might want to check out its profiling capabilities as well: `https://xdebug.org/docs/profiler`. Other commercial services that offer more comfort are—for example—**Tideways** (`https://tideways.com`) or **Blackfire** (`https://www.blackfire.io`).

> **Xdebug alternatives**
>
> Please note that you can use other extensions for this, such as **PCOV** (`https://github.com/krakjoe/pcov`), which performs better if you only want to do code coverage reports. However, Xdebug is an incredibly useful debugger that you should know—if you do not, we encourage you to check out some tutorials about it.

How to generate code coverage reports

To demonstrate how to create a code coverage report, we will use our little demo application from the previous section of this chapter. To follow up on the example, check it out from GitHub, run `composer install`, and make sure you have Xdebug installed with `mode` set to `coverage`.

Before we start generating a report, let us see which report formats PHPUnit has to offer. It can generate reports in a variety of formats that you will most likely not need right now, such as `Clover`, `Cobertura`, `Crap4J`, or the `PHPUnit` XML format. They can get more relevant, though, when you start integrating PHPUnit with other tools.

However, we do not want to do this in this book, so we are only interested in the two most accessible formats: text and HTML. The text format can directly be printed on the command line, which is useful when you want instant results or to integrate PHPUnit in your build pipeline, while the HTML format offers more information.

For our example, we want to write both report formats into a new folder called `reports` in the project root. Although you can generate them using the many PHPUnit runtime options, we want to use the `phpunit.xml` configuration file to define what to generate upon every test run. The following code snippet shows a minimal version, reduced for readability. In our GitHub repository, you will find the full `phpunit.xml` file:

```xml
<?xml version="1.0" encoding="UTF-8"?>
<phpunit bootstrap="vendor/autoload.php">
    <testsuites>
        <testsuite name="default">
            <directory>tests</directory>
        </testsuite>
    </testsuites>
    <coverage>
        <include>
            <directory suffix=".php">src</directory>
        </include>
        <report>
            <html outputDirectory="reports/coverage" />
            <text outputFile="reports/coverage.txt" />
        </report>
    </coverage>
</phpunit>
```

Besides the basic configuration, which includes a definition of the tests folder we need for the regular test runs, we added the <coverage> node. This contains two child nodes: <includes> and <report>. It is important to specify when using the <includes> node which directory and file extension should be used for collecting code coverage information. Otherwise, PHPUnit will not generate any reports, but also will not complain about missing information. This can be quite confusing at times.

Furthermore, we need to tell PHPUnit where to write which reports. We use the <report> node for this, and as you can see, we specified both the HTML and text reports to be written into the reports folder in our project root, which will be created if it does not exist.

PHPUnit expects the configuration file to be named phpunit.xml and to be in the project root. If this has been done, you can quickly execute the generation of the reports by running the following command without any further options or arguments:

```
$ vendor/bin/phpunit
```

After the execution of the preceding command, you will find a reports folder generated in your project root. It should contain two things: first, a coverage.txt file that contains a report in text format, and second, a coverage folder that contains an HTML report.

> **Code coverage is costly**
>
> Using Xdebug to generate code coverage reports will slow down the execution of your test suite because Xdebug has to collect a lot of data, and it was not built for performance. We therefore recommend you enable Xdebug and the report generation only if necessary but keep it disabled during your regular test runs.

The text report is short, but already tells you how well your tests cover your application, as shown in the following screenshot:

```
$ cat reports/coverage.txt


Code Coverage Report:
  2022-06-21 21:02:07

 Summary:
  Classes: 50.00% (2/4)
  Methods: 50.00% (3/6)
  Lines:   50.00% (6/12)

CodeCoverageExample\MyApp
  Methods: 100.00% ( 2/ 2)   Lines: 100.00% (  3/  3)
CodeCoverageExample\MyOtherClass
  Methods:   0.00% ( 0/ 2)   Lines:  33.33% (  2/  6)
CodeCoverageExample\MyRepository
  Methods: 100.00% ( 1/ 1)   Lines: 100.00% (  1/  1)
$
```

Figure 10.2: Text code coverage report

To get more details, please open the `reports/coverage/index.html` file in your browser. It should look like this:

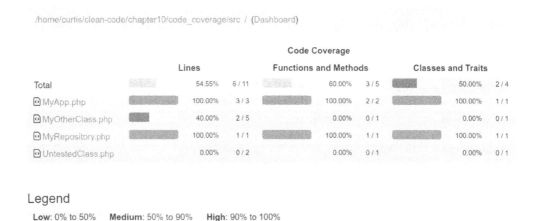

/home/curtis/clean-code/chapter10/code_coverage/src / (Dashboard)

	Code Coverage							
	Lines			Functions and Methods			Classes and Traits	
Total		54.55%	6 / 11		60.00%	3 / 5		50.00%	2 / 4
MyApp.php		100.00%	3 / 3		100.00%	2 / 2		100.00%	1 / 1
MyOtherClass.php		40.00%	2 / 5		0.00%	0 / 1		0.00%	0 / 1
MyRepository.php		100.00%	1 / 1		100.00%	1 / 1		100.00%	1 / 1
UntestedClass.php		0.00%	0 / 2		0.00%	0 / 1		0.00%	0 / 1

Legend

Low: 0% to 50%   **Medium**: 50% to 90%   **High**: 90% to 100%

Generated by php-code-coverage 9.2.15 using PHP 8.1.2 and PHPUnit 9.5.20 at Mon Jun 20 22:22:52 UTC 2022.

Figure 10.3: HTML code coverage report

You will find the same information from the text report there, but with a better visualization. Furthermore, the report is interactive. For example, if you click on the `MyOtherClass.php` link on the left, you will be taken to a detailed report for that class, as shown in the following screenshot:

Figure 10.4: HTML code coverage report – class view

Two things are noteworthy here: firstly, in the **Functions and Methods** section, you might have already recognized the CRAP metric which we introduced in *Chapter 8, Code Quality Metrics*. Here, you can finally see it in action.

Secondly, the report shows you in detail which lines have been accessed during the test (green background), and which have not (red background). If there had been any lines that were inaccessible at all (for example, another statement after the last `return` statement), it would have been displayed as **Dead Code** (yellow background). Dead code can safely be removed.

You now have a nice overview of the code coverage of your project. If files appear with a red bar, they have not been executed during the test run at all, so this is where you could improve your test suite.

Using the @covers annotation

There is one problem with code coverage: it tells you which code has been *executed* during the test, but that does not mean that the executed code has also been *tested* (that is, using assertions). This is something that PHPUnit cannot determine automatically. This means that even if your code coverage reports show 100% and green bars everywhere, it does not mean that your code is well tested. It was just executed during the run of your test suite.

To overcome this problem, it is recommended to use the @covers annotation at the class level, as illustrated here:

```
/**
 * @covers MyRepository
 */
class MyRepositoryTest extends TestCase
{
    public function testGetDataReturnsAnArray(): void
    {
        // ...
    }
}
```

This improves the accuracy of our tests because by using the @covers annotation, we explicitly declare which code is meant to be tested by our tests. For example, let us say the class under test uses an external service. You only want to test that one class, not the service it uses, so you solely write assertions that check the class under test. Without the @covers annotation, though, PHPUnit would still include the external service in the code coverage report because it was executed during the test.

You can use @covers also on a method level; however, this could cause problems if— for example— you refactor a class and extract methods to some other class. If you forget to adjust the @covers annotation on the method level here, the coverage report will not be accurate anymore.

To force usage of the @covers annotation, use the forceCoversAnnotation option in the phpunit.xml file. If it is set to true, tests that do not use the annotation will be marked as risky; they do not fail, but they appear separately in the reports as something to be improved. This way, your fellow developers (and you) will not forget to use it.

What to test

We now know how to get detailed information about how much of our code is tested. So, should you strive for complete code coverage now? Should 100% be your goal?

As we saw in the tests of our example application in a previous part of this chapter, writing tests for a class does not automatically mean that you really test every aspect of it. Here, unfortunately, even measuring the code coverage will not help. However, it can help you identify tests that do not test anything. Especially when a lot of mocks are used in a test case, it can happen that only the mock gets tested, but no "real code". Consider the following test case, which is a valid test that would pass:

```php
public function testUselessTestCase(): void
{
    $repositoryMock =
        $this->createMock(MyRepository::class);
    $repositoryMock
        ->method('getData')
        ->willReturn([
            'value_1' => 'a',
            'value_2' => 'b'
        ]);

    $this->assertEquals(
        [
            'value_1' => 'a',
            'value_2' => 'b'
        ],
        $repositoryMock->getData()
    );
}
```

This example is simplified, but it demonstrates how code coverage reports can help because this test would not add a single tested line to our code coverage ratio. Unfortunately, no tool can tell you (yet) which tests are well written and which should be improved or are even useless, as in our example.

Following the **Pareto principle**, aiming for 80% code coverage should already vastly improve your code base, and this can be achieved with a reasonable amount of effort. Put your focus on the code that makes your application special—often referred to as **business logic**. This is the code that requires most of your attention.

> **Pareto principle**
>
> The Pareto principle states that 80% of results are achieved with 20% of the total effort. The remaining 20% of the results require the quantitatively most work, with 80% of the total effort.

There is also trivial code that does not really need to be tested. A common example is testing for getters and setters. If these methods contain further logic, it makes sense to test them, of course. But if they are just simple functions that set or return the value of a property, it is a waste of time to write tests for them. Still, you would need to do this if you would like to strive for 100% code coverage.

Other examples are configuration files, factories, or route definitions. It is sufficient to use E2E or integration tests, which ensure that the application works in general. They implicitly (that is, without using concrete assertions) test all the **glue code**, which is all the code that keeps your application together but is tedious to test.

Particularly, E2E tests are usually not counted into the code coverage metric because it is technically difficult to do so. If you have them, though, they will add an extra layer of test coverage that cannot be measured. You cannot brag about 100% code coverage, but you know that all the different test types have your back, and that should be our number one goal.

Summary

In this chapter, we discussed why you should use automated testing and how it improves your code quality. We covered the main three testing types, which are unit testing, integration testing, and E2E testing, with their pros and cons, potential pitfalls, and our recommendations on how to use them. Finally, you learned about the concept of code coverage, and how to use it in your own projects.

Together with the knowledge from the previous chapter about code quality tools and how to organize them, in the next chapter, we can finally start combining all these tools together into a process that helps to run all of them in structured and reliable ways—build pipelines.

Further reading

There are more test types out there than we could cover in this chapter. If you find the world of automated testing as fascinating as the authors do, you might want to check out other test types as well, such as the following:

- **Mutation testing** is about modifying the code to be tested in tiny changes (so-called mutants). If your tests can catch these mutants, they are usually well written; otherwise, they will let the mutant escape. **Infection** is currently the best-known tool for this test type in the PHP world (`https://infection.github.io`).

- **Visual regression testing** literally compares screenshots of the application made during tests with original screenshots to catch problems in **Cascading Style Sheets** (**CSS**). While this is not directly PHP-related, it could be interesting for you if you want to keep the styling of your web project perfect. A good candidate to check is **BackstopJS** (`https://github.com/garris/BackstopJS`).

- **API testing** can be considered E2E testing, but just for the API your application might provide. Since the tests are based on **Hypertext Transfer Protocol** (**HTTP**) requests, a headless browser is not needed, which makes the setup easier. A good choice to start with API testing is **Codeception** (`https://codeception.com`).

- **Behavior-driven development** (**BDD**) is a very interesting approach because it focuses on the communication between stakeholders (for example, the project manager), QA (if there is any), and the developers. This is done by a special way of writing tests in a language called **Gherkin**, which basically enables non-technical people to write test suites. The BDD tool for PHP is called **Behat** (`https://github.com/Behat/Behat`).

11

Continuous Integration

You have learned the theory about writing clean **PHP: Hypertext Preprocessor** (**PHP**) code, and you now know the necessary tools and metrics that help us to achieve and keep to high quality levels. However, what is still missing is the integration of all these technologies into a workflow that facilitates your daily work.

In the following pages, we will elaborate on **continuous integration** (**CI**) and learn by example how to set up a simple but effective automated workflow.

Furthermore, we will show you how to set up a selection of code quality tools locally in a way that they support you the most, without having to manually run them. Additionally, we will tell you some best practices about how to add these workflows to an existing project.

The main topics we will cover are listed here:

- Why you need CI
- The build pipeline
- Building a pipeline with GitHub Actions
- Your local pipeline—Git hooks
- Excursion—Adding CI to existing software
- An outlook on **continuous delivery** (**CD**)

Technical requirements

Additional to the setup of the previous chapters, you will require a GitHub account to be able to follow all examples. This will come with no additional costs, though, as we are only using the free plan.

The example application that we will use in this chapter can be downloaded from the GitHub repository to this book: `https://github.com/PacktPublishing/Clean-Code-in-PHP/tree/main/ch11/example-application`.

Why you need CI

Writing software is a time-consuming and thus costly process. If you develop software for fun, it will "only" cost you your leisure time. If you work for a company (be it as a contractor or full-time employee), the time is even more valuable, as you get paid for it. As a matter of fact, companies want to reduce costs, and thus they do not want to spend more money on a feature than necessary.

A big part of our daily work is to fix defects. Delivering bug-free software is something that probably all developers would like to achieve. We do not make mistakes deliberately, yet they will always happen. There are ways, however, to reduce the costs of bugs.

The costs of a bug

A bug is considerably costly because it adds no value to the product. Therefore, we aim to catch these bugs as early as possible—the earlier we catch them, the fewer costs they will cause. The following screenshot visualizes how the costs to fix a bug increase significantly the later it appears in the development process:

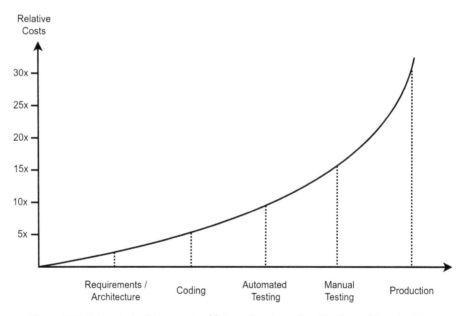

Figure 11.1: Estimated relative costs of fixing a bug based on the time of its detection

But what is the reason for the massive cost increase over time? And why do bugs even cost money?

In the early stages, costs mainly arise from the required time to resolve the issue. If a bug could have been avoided just by better requirements, for example, then this requires less effort as it was discovered during the manual testing. If a bug was found in production, many people are involved in fixing it: first, a helpdesk employee needs to acknowledge the bug reported by a customer and pass it over to the **quality assurance (QA)** engineer, who reproduces it and writes a proper bug ticket.

This ticket then gets assigned to the product manager, who, after taking time to reproduce and verify the defect as well, plans it in for the next sprint. The ticket eventually gets assigned to one developer, who will need some time to reproduce and fix it. But it is not over here because the bug fix probably requires a code review from another developer and gets double-checked from the product manager or QA engineer as well before it can finally be released.

Once "escaped" from the developer's local environment, all this overhead raises the costs of the defect significantly. Additionally, if the bug is already in production, it can lead to customers not wanting to use the product anymore because they are no longer happy with it. This is called **customer churn**.

Even if you are not working on a commercial product, but—for example—on an open source project, the concept can be translated into time or effort. A bug will lead to an issue report that you first need to read and understand, probably asking some more questions, and waiting for the ticker author to reply. If your software is too buggy, people will use it less, and all your previous efforts might have been in vain at some point.

How to prevent bugs

Fortunately, we now have a whole toolbox at your side that can help us find bugs in your code before somebody else does. We just have to use it—and this is already a problem because we developers are usually lazy people.

Of course, you could run all the tools manually before every deploy. Let us say you want to deploy some code to production. After merging the code into the `main` branch, the following steps should be executed to ensure that no broken code gets delivered to production:

1. Using the PHP linter to ensure the syntactical correctness of the code
2. Executing a code style checker and fixer to keep the code styling aligned
3. Finding potential issues using static code analysis
4. Execution of all automated test suites to ensure your code still works
5. Creating reports for the used code quality metrics
6. Cleaning up the build folder and creating an archive of the code to deploy

This is quite a list of things to keep in mind. Nobody would do this over a longer period without making mistakes at some point, so, naturally, you would start writing scripts that help you to execute these steps in one go. This is already a good improvement, and we will also make use of it a bit further on in this chapter.

Introducing CI

Running all the steps from the previous section in your local environment will take some time, and during the checks run, you can hardly work on anything else, so you have to wait until they are finished. So, why not load this whole workflow onto another, dedicated server?

This is precisely what CI does: it describes the automated process of putting all necessary components of your application together into a **deliverable** so that it can be deployed to the wanted environments. During the process, automated checks will ensure the overall quality of the code. It is important to keep in mind that if one of the checks fails, the whole build will be considered as failed.

There are many **CI tools** available, such as Jenkins, which is usually self-hosted (that is, operated by you or someone in your team or company). Or, you can choose paid services such as GitHub Actions, GitLab CI, Bitbucket Pipelines, or CircleCI.

You will often read the abbreviation *CI/CD*, and we will also use it throughout this book. **CD** stands for **continuous delivery**, a concept that we will cover at the end of this chapter. For now, you do not need to care about it, though.

Setting up one of these tools sounds like a lot of work, but it also has some great benefits, such as the following:

- **Scalability**: If you work in a team, using the local setup will quickly cause problems. Any changes to the build process would need to be done on the computer of every developer. Although the build script would be part of your repository, people may forget to pull the latest changes from it before deploying, or something else might go wrong.

- **Speed**: Automated tests or static code analysis is a pretty resource-consuming job. Although today's computers are powerful, they have to do a lot of concurring tasks, and you do not want to additionally run a build pipeline on your local system. **CI/CD servers** are doing only this one job, and they are usually doing it fast. And even if they are slow, they still take the load from your local system.

- **Non-blocking**: You need a build environment to run all the tools and checks on your code. Using your local development environment for this will simply block it for the duration of the build, especially when you use slower test types such as integration or **end-to-end** (E2E) tests. Running two environments on your local system—one for development and one for CI/CD—is not recommended, as you will quickly end up in a configuration hell (just think of blocking a database of web server ports).

- **Monitoring**: Using a dedicated CI/CD server will let you keep an overview of who deployed what and when. Imagine that your production system is suddenly broken—using a CI/CD server, you can immediately see what the last changes have been and deploy the previous version of your application with a few clicks. Furthermore, CI/CD tools keep you up to date and inform you—for example—via email or your favorite messenger application about any build and deploy activities.

- **Handling**: A handwritten deployment script will surely do the work, but it takes a lot of time to make it as comfortable and flexible as a modern CI/CD solution. Plus, if you go with the business standards, it is much more likely that other developers of your team will already have experience with it.

The preceding points hopefully give you an idea of how much you will benefit from using CI. An integral part of each CI/CD tool is the so-called build pipeline, which we will explain in detail in the next section.

The build pipeline

In the previous section, we listed the many necessary steps to make our code ready to be shipped to production. In the context of CI, a combination of these steps is what we call the **build pipeline**: it takes the input (in our case, all the application code), runs it through several tools, and creates so-called **build artifacts** out of it. They are the outcome of a build—usually, this includes the **deliverable** (a package of the application code that is ready to be moved to the desired environment), plus additional data, such as build logs, reports, and so on.

The following diagram gives you a schematic overview of what a typical build pipeline could look like. Since it is not executed in your local environment, it requires two additional steps: *creating a build environment* and *building an application*:

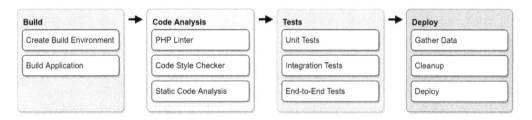

Figure 11.2: Schema of a CI pipeline

> **Other languages in the pipeline**
>
> In this book, we will only have a look at the PHP-related parts of the pipeline, yet a modern web application does not only consist of PHP. Especially for the frontend-facing code, there is another whole universe of tools that needs to be part of the pipeline, too. Yet in the end, the process is very similar.

In the next sections, we will go into more detail about every build stage. We will keep it theoretical at first, and then give examples of technical implementation later in this chapter.

Stage 1: Build project

The CI pipeline requires a dedicated build instance of our application, where we can run all tools and checks on it isolated. This can roughly be divided into two steps: creating a build environment and running the necessary build tools.

Creating a build environment

To build an application somewhere else than on the local development system, we first need to create a **build environment**. How exactly the environment is provided depends on the used CI/CD tool. This can either be a dedicated server that offers separated workspaces for every project, or a completely containerized Docker environment, which is spun up every time you require it and only lasts for the duration of the build.

Once a build environment exists, we need to download all the source code there, but without external packages or other dependencies for now. Most likely, your code will be stored in a Git repository and either hosted on a private Git server, or a commercial service. Downloading a copy of a specific branch of a repository is called **checkout**.

We have to pay attention to which branch of the repository the code gets checked out from. This depends on what you want to build. If you intend to check the code of a **pull request** (**PR**) (a set of changes to the code, which someone requests to be integrated into the code base), then you need to check out the branch of that feature. If you wish to upload the latest version of your application to production, you want to check out the `main` branch.

> **Main versus master branch**
>
> Throughout the history of computers, the terms *master* and *slave* have been widely used, be it for hard disk configuration, database replication, or—of course—Git. However, these terms are harmful to many people, so the **information technology** (**IT**) people decided to finally change this into less offensive terms, such as *main* and *replica*. GitHub, for example, by default calls the `main` branch simply *main* instead of *master* nowadays. You will still find repositories out there that use the old branch name. In this book, though, we will stick with the new terminology.

Do not worry, however, if your project is not hosted using Git—this step is still necessary because we need to get the code on the CI/CD server. Be it via Git, Mercurial, **Subversion** (**SVN**), or even direct file download, it does not matter in the end. The outcome of this step is to have the code we want to deploy readily available on the CI/CD server so that we can start installing the dependencies in the next step.

Building the application

Building the application is similar to installing it on a new system. In the previous step, we made sure that the source code is available in the environment. In this step, we need to execute any required build steps. This usually includes the following:

- **Installing external dependencies**: Your repository should only contain your own code, with no external dependencies. These we manage, for example, via Composer or the **PHAR Installation and Verification Environment** (**Phive**).

- **Creating configuration files**: Your repository should not contain any passwords or other confidential data such as **application programming interface (API)** keys. They can be securely stored in the CI/CD tool and used to create an environment file (.env, for example) during this stage.

- **Preparing the test database**: To run integration or E2E tests, the build instance needs a working database. Commonly, this is done by creating a test database, importing the database schema, running any additional database migrations, and—finally—populating the database with test data.

To reduce the build time, many modern CI/CD tools offer caching. If activated, they will keep the dependencies in temporary storage after the first download. It is generally a good idea to turn this on, if not activated by default.

Stage 2 – Code analysis

We covered code quality tools in *Chapter 7, Code Quality Tools,* in detail. Now, it is time to add these tools to our pipeline to ensure that they will be executed for every change that gets introduced.

PHP linter

If you merge code into another branch, it can always happen that the code breaks. Git has very sophisticated merge algorithms, but still, it cannot do magic. Given that you have a test suite with high coverage, some tests will surely break if there were a syntax error that had been caused by the merge. So, why should we run this extra step? We recommend it because the PHP linter has two advantages: it runs very fast, and it will check all PHP files, no matter if there are tests for them or not.

We want our pipeline to fail fast if any issues have been detected. Therefore, before you execute any long-running tasks, it makes sense to run a quick syntax check in the beginning. They would break in any case, and you would lose some valuable time. As a rule of thumb, the faster the check runs, the earlier it will appear in the pipeline.

Code style checker

After checking the code syntax, it is time to check the code style. This operation is fast too, so it makes sense to run it early in the pipeline. For our example, we will go with the **PHP Coding Standards Fixer (PHP-CS-Fixer)**, which we already introduced in *Chapter 7, Code Quality Tools.*

There is a subtle but important difference between running PHP-CS-Fixer locally and in the CI/CD pipeline: for the latter, we will only use it to check the code, but not to fix it. We do not want the pipeline to *change* our code, just to *analyze* it. In other words, our pipeline will only check if the code is correctly formatted (according to the rules we defined), but it will not attempt to fix it automatically; if any of the rules are violated, the build has failed.

No rules are saying that the CI/CD pipeline should not change the code. However, it adds complexity to automatically commit the changes to the repository during the process. Moreover, it requires a very well-tested application, and you need to trust that the tool of your choice does not break anything. Usually, they work well, but do you want to risk it?

In your local environment, it makes sense to run the fixer alongside the code style checker. We will discuss the local setup in one of the next sections of this chapter.

Static code analysis

At this point, we know that our code is syntactically correct and formatted according to our rules. Both previous checks are usually performed quickly, so our build would fail early if any of those easy-to-detect problems had occurred.

Now, it is time to run the slower tasks. The static code analysis usually takes a bit longer than the two previous ones, but it is by far not as slow as running the automated tests. Essentially, this step is not that different from the linting and code style checking: if the rules we have defined earlier are violated, the build will fail.

If you are introducing CI to an existing project, the challenge is to find the sweet spot in error reporting. On the one hand, you want to keep the developers happy and not force them to fix dozens of issues introduced by other developers on every file they touch. On the other hand, you need to set the threshold tight enough to enforce at least some refactoring with every code change.

There is no golden rule here, unfortunately, and you need to experiment with the settings. At a later point, when most of the issues of the static code analysis reports are solved, you need to tighten the error reporting rules a little so that your project does not stagnate at a certain level.

Stage 3 – Tests.

Once our code has reached this point in the pipeline, we are convinced that it is syntactically correct, adheres to our code styling guidelines, and has no general flaws according to our static code analysis rules. So, we will now run the step in the pipeline, which usually takes the longest time: the automated tests.

As we introduced in the previous chapter, there are more than just unit tests to consider. Often, a project such as a web service has at least some integration tests to ensure that the service is working fine as a whole, including the database transactions. Or, if your project is a classical web application, you might have an E2E test suite that utilizes a browser to virtually click through it.

We want to apply the same approach here that we did for the build steps: start with the fast-running tests, and then continue with the slower ones. If the unit tests are failing already, you do not need to wait for the results of the E2E tests. So, for the tests, the execution order would usually be this:

1. Unit tests
2. Integration tests
3. E2E tests

If only one test of whatever type fails, the build will be marked as failed. If they all pass, we have already passed the most critical part of the pipeline. Our application is ready to be deployed, so now, it is time to clean up and prepare the deliverable.

Stage 4 – Deploy

Our code has been checked thoroughly by various tools, and we are confident that it adheres to our standards. We can now prepare the **build artifacts** and finally deploy the application. Let us have a look at what needs to be done for this.

Gathering data

During the previous stages, all tools we used produced some sort of data, be it through writing to the **standard output** (**stdout**), or if you configured that, by creating reports that summarize the performed actions.

You can, for example, upload generated reports to dedicated storage or push them to your repository. Or, you can use the code coverage report of PHPUnit and automatically create a code coverage badge from it, which you can add to the README of your GitHub project.

The most important use case, though, is to debug if any of the stages have failed. You can never have enough debug output in case something goes wrong, so it is a good idea to set the verbosity of your tools to a higher level. Your CI/CD tool usually takes care that everything that was written to the stdout is made available after the execution of the build pipeline.

Cleanup

Before we upload the application somewhere, we want to make sure that it does not contain any unnecessary ballast. This includes removing logs or reports from previous stages or deleting the code quality tools. Remember, we should only deploy code to production that is necessary for the application to run—development tools such as PHPUnit are not built with security in mind (`https://phpunit.readthedocs.io/en/9.5/installation.html#webserver`).

Deploy

To deploy the code to the target environment, we need to wrap it up into an asset that can be easily moved there. This asset is also called a **deliverable**. The type we choose for the deliverable depends on how the application has to be deployed to production. A common type for such a deliverable is simply an archive of the code that has to be deployed.

For example, if your production environment is running on a classic on-premises web server, we need to create an archive of the application code, upload it to the target server, and extract it from there.

The de facto standard today is a containerized environment using Docker. Once the build instance has been thoroughly tested, a **Docker image** can be created from it. This image will then be uploaded to an **image repository** such as **Amazon Web Services Elastic Container Repository** (**AWS ECR**). Such an **image repository** hosts all your images, so they can be used to spin up new containers when needed. This approach paved the way for highly scalable web applications as we have them today, so designing your application to be **Dockerizable** from the beginning will pay off if your application, at some point, needs to scale.

> **Containers and images**
>
> If you are new to Docker, the concept of containers and images can be confusing. In short, an image contains all the data, but is read-only and cannot be used alone. To make it usable in the Docker environment, it requires a container that contains all information from the image and offers the required functionality to connect to other containers of your setup. Since you can create as many containers as you need from an image, you can also think of it as a container template.
>
> If you want to know more about Docker, we recommend you check the official documentation at `https://docs.docker.com/get-started/overview`. There are also tons of tutorials, online courses, and other information to be found on the internet.

We now have a good idea of what a build pipeline could look like. Before we start setting up our example pipeline, one more thing needs to be clarified—when will it be executed?

Integrating the pipeline into your workflow

After setting up all the necessary steps, we finally need to integrate the pipeline into your workflow. CI/CD tools usually offer you different options on when a pipeline gets executed. In the beginning, this can, of course, be done manually by clicking a button. This is not very comfortable, though. If you use hosted Git repositories such as GitHub, GitLab, or Bitbucket, you can connect them with your build pipeline and start the build whenever a PR was created or a branch got merged into the `main` branch.

For huge projects where the build takes hours, it is also common to run a build for the current code base at nighttime (so-called **nightly builds**). The developers then get their feedback from the pipeline the next day.

Running a build requires some time, and of course, the developers should not sit in front of the screens and wait until they can continue with their work. They should rather be informed as soon as the build succeeded or failed. All CI/CD tools nowadays offer multiple ways to notify the developer, mostly by email and by messages in chat tools such as Slack or Microsoft Teams. Additionally, they often also offer dashboard views, where you can see the status of all the builds on one screen.

You now should have a good idea of what a build pipeline could look like for your project. Therefore, it is time to show you a practical example.

Building a pipeline with GitHub Actions

After learning about all the stages of CI, it is time to practice. Introducing you to all the features of one or more CI/CD tools is out of scope for this book; however, we still want to show you how easy it can be to set up a working build pipeline. To keep the barrier to entry as low as possible for you and to avoid any costs, we have decided to use **GitHub Actions**.

GitHub Actions is not a classic CI tool like Jenkins or CircleCI, but rather a way to build workflows around GitHub repositories. With a bit of creativity, you can do much more than "just" a classical CI/CD pipeline. We will only focus on that aspect, of course.

You probably already have a GitHub account, and if not, getting one will not cost you anything. You can use GitHub Actions for free up to 2,000 minutes per month at the time of writing for public repositories, which makes it a great playground or a useful tool for your open source projects.

> **Example project**
>
> We created a small demo application to use during this chapter. You will find it here: `https://github.com/PacktPublishing/Clean-Code-in-PHP/tree/main/ch11/example-application`. Please note that it does not serve any other purpose than demonstrating the basic use of GitHub Actions and Git hooks.

GitHub Actions in a nutshell

GitHub Actions offers no fancy user interface where you can configure all stages. Instead, everything is configured via **YAML Ain't Markup Language** (**YAML**) files that are stored directly in the repository. As a PHP developer, you are most likely experienced with using YAML files for all sorts of configurations—if not, do not worry, as they are easy to understand and use.

GitHub actions are organized around workflows. A workflow gets triggered by certain events and contains one or more jobs to be executed if the event occurred. A job consists of one or more steps that execute one action each.

The files have to be stored in the `.github/workflows` folder of a repository. Let us have a look at the first lines of the `ci.yml` file, which will be our CI workflow:

```yaml
name: Continuous Integration
on:
  workflow_dispatch:
  push:
    branches:
      - main
  pull_request:
jobs:
  pipeline:
    runs-on: ubuntu-latest
    steps:
      - name: ...
        uses: ...
```

That is quite some information already. Let us go through it line by line:

- `name` defines how the workflow will be labeled within GitHub and can be any string

- `on` states which events should trigger this workflow; these comprise the following:

  - `workflow_dispatch` allows us to manually trigger the workflow from the GitHub website, which is great for creating and testing a workflow. Otherwise, we would need to push a commit to main, or create a PR every time.

  - `push` tells GitHub to execute this workflow whenever a push happens. We narrow it down to pushes on the `main` branch only.

  - `pull_request` will additionally trigger the workflow on every new PR. The configuration might look a bit incomplete because there is no more information after the colon.

- `jobs` contains a list of jobs to be executed for this workflow, as detailed here:

  - `pipeline` is the **identifier (ID)** of the only job in this YAML. We chose the word *pipeline* to illustrate that we can use GitHub Actions to build our CI/CD pipeline. Note that an ID must consist of one word or several words, concatenated by an underscore (`_`) or a dash (`-`).

  - `runs-on` tells GitHub to use the latest Ubuntu version as a runner (that is, as a platform) for this job. Other available platforms are Windows and macOS.

  - `steps` marks a list of steps to be executed for this job. In the next section, we will have a closer look at this.

We have the basics of workflow configured now, so we can begin adding the build stages.

Stage 1 – Build project

The steps are what makes GitHub Actions so powerful: here, you can choose from a vast amount of already existing *actions* to use in your workflow. They are organized in *GitHub Marketplace* (`https://github.com/marketplace`). Let us add some steps to the workflow YAML, as follows:

```
steps:
  ###################
  # Stage 1 - Build #
  ###################
  - name: Checkout latest revision
    uses: actions/checkout@v3
  - name: Install PHP
    uses: shivammathur/setup-php@v2
    with:
```

```
php-version: '8.1'
coverage: pcov
```

Actions maintained by GitHub are to be found in the `actions` namespace. In our example, this is `actions/checkout`, which is used to check out the repository. We do not need to specify any parameters for now, as this action will automatically use the repository in which this workflow file is located.

The `@V3` annotation is used to specify the *major* version to use. For `actions/checkout`, this would be version 3. Please note that the latest *minor* version is always used, which at the time of writing would be version `3.0.2`.

The other action, `shivammathur/setup-php`, is provided by one of the many great people who make their work available as open source. For this step, we are using the `with` keyword to specify further parameters. In this example, we use the `php-version` option to have PHP `8.1` installed on the previously selected *Ubuntu* machine. Using the `coverage` parameter, we can tell `setup-php` to enable the `pcov` extension for generating code coverage reports.

> **Action parameters**
>
> Both actions introduced previously offer far more parameters than we can describe here. You can find more information about their functionality by looking them up in *Marketplace*.

Regarding the formatting, we used comments and blank lines between the steps to make the file more readable. There is no convention, and it is completely up to you how to format your YAML files later.

The next step is the installation of the project dependencies. For PHP, this usually means running `composer install`. Please note that we *do not* use the `--no-dev` option because we need to install the `dev` dependencies to perform all the quality checks. We will remove them at the end of the pipeline again.

> **Dependency management**
>
> We use the **Composer** workflow in this chapter as an example to manage our code quality tools, because this is the most common way. However, both other ways of organizing the code quality tools we introduced in *Chapter 9, Organizing PHP Quality Tools,* would work with GitHub Actions as well. In that chapter, we also explained the `--no-dev` option in detail.

This is what the next steps could look like:

```
- name: Get composer cache directory
  id: composer-cache
  run: echo "::set-output name=dir::$(composer config cache
```

```
      files-dir)"
  - name: Cache dependencies
    uses: actions/cache@v2
    with:
      path: ${{ steps.composer-cache.outputs.dir }}
      key: ${{ runner.os }}-composer-${{
        hashFiles('**/composer.lock') }}
      restore-keys: ${{ runner.os }}-composer-
  - name: Install composer dependencies
    run: composer install
```

GitHub actions require some manual work to make caching of Composer dependencies possible. In the first step, we store the location of the Composer cache directory, which we get from Composer using the `config cache-files-dir` command, in an output variable called `dir`. Note `id: composer-cache` here—we will need this to reference the variable in the next step.

Then, we access this variable in the next step by using the `steps.composer-cache.outputs.dir` reference (a combination of the `id` value we set in the previous step, and the variable name) to define the directory that should be cached by the `actions/cache` action. `key` and `restore-key` are used to generate unique caching keys—that is, the cache entries where our Composer dependencies are stored.

Lastly, we use the `run` parameter to directly execute `composer install`, as if we would execute it locally on an Ubuntu machine. This is important to keep in mind: you can, but you do not have to use existing GitHub actions for every step—you can just execute pure shell commands (or equivalent commands on Windows runners) as well.

There are also actions in *Marketplace* that take over the writing of commands, such as `php-actions/composer`. We do not have a preferred solution here; both will work fine.

Because we want to run integration tests on the API of our example application, we need to have a web server running. For our simple use case, it is totally enough to use the PHP built-in web server, which we can start using in the following step:

```
  - name: Start PHP built-in webserver
    run: php -S localhost:8000 -t public &
```

The `-S` option tells the PHP binary to start a web server that is listening on the `localhost` address and port `8000`. Since we start in the `root` folder of our project, we need to define a *document root* folder (the folder where the web server looks for files to execute) using the `-t` option. Here, we want to use the public folder, which only contains the `index.php` file. It is good practice to not store any other code in the document root folder since this makes it harder for attackers to hack our application.

> **PHP built-in web server**
>
> Please note that the built-in web server of PHP is only to be used for *development* purposes. It should never be used in *production*, since it was not built with performance or security in mind.

You surely noticed the ampersand (&) at the end of the command. This tells Linux to execute the command, but not wait for its termination. Without it, our workflow would get stuck at this point because the web server does not terminate by itself, as it needs to keep listening for requests until we run our *integration tests* at a later stage.

The setup of our build environment is complete. Now, it is time to run the first code quality checks on our sample application.

Stage 2 – Code analysis

In the first build stage, we created our build environment and checked out our application code. At this point, the application should be completely functional and ready to be tested. Now, we want to do some static code analysis.

The standard approach is to use dedicated GitHub actions for each tool. The benefit is that we keep the development tools away from the build environment, as they will be executed in separate Docker containers that will be discarded right after use. There are some drawbacks to this approach, though.

Firstly, with each action, we introduce yet another dependency, and we rely on the author to keep it up to date and not lose interest in maintaining it after a while. Additionally, we add some overhead, since Docker images are usually many times bigger than the actual tool. And lastly, when our application setup gets more complicated, running the code quality tools in separate Docker containers can cause issues, simply because it is not the same environment as the build environment. Sometimes, already tiny differences can cause problems that keep you engaged for hours or days in solving them.

As we saw in the previous section, we can simply execute Linux shell commands in the build environment, so nothing speaks against executing our code quality tools directly on the build environment—we just need to make sure to remove them afterward so that they do not get released into production.

In our example application, we added PHP-CS-Fixer and PHPStan to the `require-dev` section of the `composer.json` file. By adding the following lines of code to our workflow YAML, we will let them execute as the next steps:

```
###########################
# Stage 2 - Code Analysis #
###########################
- name: Code Style Fixer
  run: vendor/bin/php-cs-fixer fix --dry-run
```

```
- name: Static Code Analysis
  run: vendor/bin/phpstan
```

We do not need many parameters or options here, since our example application provides both the
`.php-cs-fixer.dist.php` and `phpstan.neon` configuration files, which both tools will look
up by default. Only for PHP-CS-Fixer will we use the `--dry-run` option because we only want to
check for issues during the CI/CD pipeline, and not solve them.

> **Setting the scope of checks**
>
> For our small example application, it is OK to run the preceding checks on all files because
> they will execute quickly. If our application grows, however, or we wish to introduce CI/CD
> to an existing application (which we will discuss further on in this chapter), it is sufficient to
> run these checks on those files that have only changed in the latest commit. The following
> action could be helpful for you in this case: `https://github.com/marketplace/`
> `actions/changed-files`.

If neither PHP-CS-Fixer nor PHPStan reports any issues, we can safely execute the automated tests
in the next stage: the tests.

Stage 3 – Tests

Our code has been thoroughly analyzed and checked for bugs and syntax errors, yet we need to
check for logical errors in our code. Luckily, we have some automated tests to ensure that we did not
inadvertently introduce any bugs.

For the same reasons as for the code quality tools in *Stage 2*, we do not want to use a dedicated action for
running our PHPUnit test suites. We simply execute PHPUnit as we would on our local development
system. Using the `phpunit.xml` file clearly proves useful here since we do not need to remember
all the many options to use here. Let us have a look at the *workflow YAML* first, as follows:

```
##################
# Stage 3 - Tests #
##################
- name: Unit Tests
  run: vendor/bin/phpunit --testsuite Unit
- name: Integration Tests
  run: vendor/bin/phpunit --testsuite Api
```

The only thing worth noting here is that we do not just run all tests, but we split them up in two test
suites: `Unit` and `Api`. Since our unit tests should execute the fastest, we want to run them (and fail)
first, then followed by the slower integration tests. Please note that we did not add any E2E tests as
our application does not run in the browser but is a mere web service.

We split the tests up by using the `phpunit.xml` configuration file. The following code fragment shows you its `<testsuites>` node, where we separate the suites by their directory (`Api` and `Unit`):

```
<testsuites>
    <testsuite name="Api">
        <directory>tests/Api</directory>
    </testsuite>
    <testsuite name="Unit">
        <directory>tests/Unit</directory>
    </testsuite>
</testsuites>
```

We also configured PHPUnit to create code coverage reports, as illustrated here:

```
<coverage processUncoveredFiles="false">
    <include>
        <directory suffix=".php">src</directory>
    </include>
    <report>
        <html outputDirectory="reports/coverage" />
        <text outputFile="reports/coverage.txt" />
    </report>
</coverage>
```

To create these reports, PHPUnit will automatically use the `pcov` extension, which we configured in *Stage 1*. They will be written into the `reports` folder, which we will take care of in the next stage.

That is already everything that needs to be done for the tests stage. If our tests did not discover any errors, we are good to go into the last stage of our pipeline and wrap everything up.

Stage 4 – Deploy

Our application is now thoroughly checked and tested. Before we are ready to deploy it into whichever environment we envisioned, we need to take care of removing the `dev` dependencies first. Luckily, this is very easy, as we can see here:

```
####################
# Stage 4 - Deploy #
####################
- name: Remove dev dependencies
  run: composer install --no-dev --optimize-autoloader
```

Running `composer install --no-dev` will simply delete all the `dev` dependencies from the `vendor` folder. Another noteworthy feature is the `--optimize-autoloader` option of Composer: since in production, we will not add or change any classes or namespaces as we would do in development, the Composer autoloader can be optimized by not checking for any changes, and thus disk access, speeding it up a bit.

As the very last step, we want to create build artifacts: one artifact is the deliverable—that is the code we intend to deploy. The other artifact is the code coverage reports we created in *Stage 3*. GitHub Actions will not keep any additional data than the logging information displayed on the GitHub website after the workflow YAML has been executed, so we need to make sure they are stored away at the end. The code is illustrated in the following snippet:

```
- name: Create release artifact
  uses: actions/upload-artifact@v2
  with:
    name: release
    path: |
      public/
      src/
      vendor/
- name: Create reports artifact
  uses: actions/upload-artifact@v2
  with:
    name: reports
    path: reports/
```

We use the `actions/upload-artifacts` action to create two ZIP archives (called *artifacts* here): `release` and `reports`. The first contains all files and directories we need to run our application on production, and nothing more. We omit all the configuration files in the root folder of our project, even the `composer.json` and `composer.lock` files. We do not need them anymore, since our vendor folder already exists.

The `reports` artifact will just contain the `reports` folder. After the build, you can simply download both ZIP archives separately on GitHub. More about this in the next section.

Integrating the pipeline into your workflow

After adding the workflow YAML to the `.github/workflows` folder (for example, `.github/workflows/ci.yml`), you only need to commit and push it to the repository. We configured our pipeline to run upon every opened PR or whenever someone pushes a commit to the `main` branch.

When you open `https://github.com` and go to your repository page, you will find an overview of your last workflow runs on the **Actions** tab, as shown in the following screenshot:

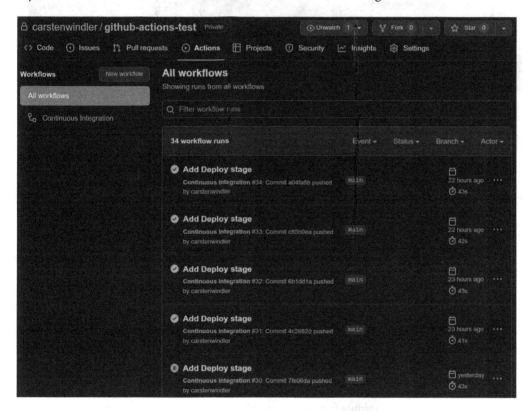

Figure 11.3: The repository page on github.com

The green checkmark marks the successful runs, while the red cross—of course—marks the failed ones. You can also see when they were executed and how long this took. By clicking on the three dots on the right side of each entry, you will find more options—for example, where you can delete the workflow run. Clicking on the title of the run, which is the corresponding commit message of the run, you will enter the **Summary** page, as shown in the following screenshot:

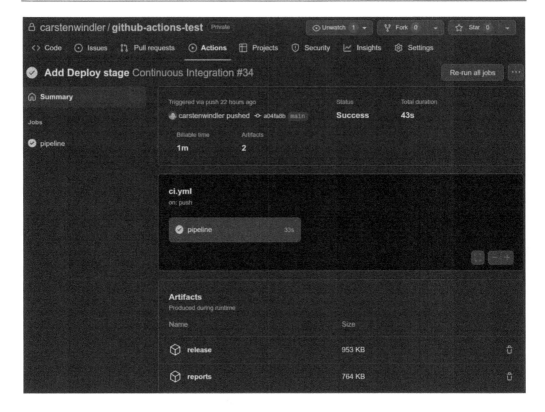

Figure 11.4: The workflow run summary page

Here, you can see all jobs of the workflow. Since our example only consists of one job, `pipeline`, you will only see one. On this page, you can also find any generated artifacts (such as our release and reports artifacts) and download or delete them. GitHub offers only limited disk space for free, so make sure to delete them when you are running out of space.

Another important piece of information is the billed time. Although our job only ran for 43 seconds in total, GitHub will deduct 1 minute from your monthly usage. GitHub offers a generous free plan, but you should have a look at your usage from time to time. You can find more information about this on your user settings page in the **Billing and plans** section (`https://github.com/settings/billing`).

If you want to see what exactly happens during the workflow run—for example, if something goes wrong—you can click on the `pipeline` job to get a detailed overview of all of its steps, as illustrated in the following screenshot:

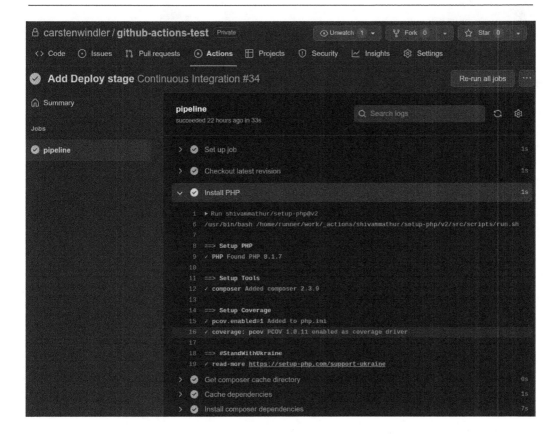

Figure 11.5: Job details page

Each step can be expanded and collapsed to get additional information about what exactly happened during its execution. In the preceding screenshot, we expanded the **Install PHP** step to see what the action did in detail.

Congratulations—you now have a working CI pipeline for your project! This ends our little tour through GitHub Actions. Of course, you can extend the pipeline as you like—for example, by uploading the release artifact to an **SSH File Transfer Protocol** (**SFTP**) server or an AWS **Simple Storage Service** (**S3**) bucket. There is a lot more than can be done, so make sure to experiment with it.

In the next section, we will show you how you can set up your local pipeline. This will save you some time and probably even costs by avoiding unnecessary workflow runs through early checks.

Your local pipeline – Git hooks

After we successfully set up a simple but already very useful CI/CD pipeline, we now want to look at running some steps already in the local development environment, even before committing them to the repository. This may sound like double work right now—why should we run the same tools twice?

Remember *Figure 11.1–Estimated relative costs of fixing a bug based on the time of its detection*, from the beginning of the chapter: the earlier we catch a bug, the fewer costs or the less effort it will cause. Of course, if we find a bug during the CI/CD pipeline, that is still much earlier than in the production environment.

The pipeline does not come for free, though. Our example application build was fast and just took roughly a minute. Imagine, however, a full-fledged Docker setup that already takes a considerable amount of time to create all the necessary containers. And now, it fails to build just because of a little bug that you could have solved within 2 minutes if you had not forgotten to execute unit tests before committing your code. You might have just taken a well-deserved tea or coffee break, only to find the build failed because of that when you came back. This is annoying and a waste of money as well as computational power.

Exactly those fast-running checks such as unit tests, a code sniffer, or static code analysis are what we want to execute before we start a full build for our changes. We cannot rely on ourselves to execute these checks automatically because we are humans. We forget things, but machines do not.

If you use Git for your development, which most developers do today, we can utilize the built-in functionality of Git hooks to automate these checks. Git hooks are shell scripts that are automatically executed on certain events, such as before or after every commit.

For our needs, the *pre-commit* hook is particularly useful. It will be executed each time you run the `git commit` command and can abort the commit if the executed script returned an error. In this case, no code would be added to the repository.

Setting up Git hooks

Setting up Git hooks manually does require some knowledge of shell scripting, so we want to use a package called `CaptainHook` to assist us. Using this tool, we can install any hook we like and even use some advanced features without the need to master Linux.

You can easily download the Phar by using Phive (see *Chapter 9, Organizing PHP Quality Tools*, for more information on that), or use Composer to install it, as we will do now:

```
$ composer require --dev captainhook/captainhook
```

Next, we need to create a `captainhook.json` file. This file contains the hook configuration for your project. Since this file will be added to the repository, we ensure that other developers in our team can use it. To create this file, we could run the following command:

```
$ vendor/bin/captainhook configure
```

CaptainHook will ask you a couple of questions and generate a configuration file based on your answers. You can skip this step, however, and create a file directly, as we will do now. Open your favorite editor, and write the following code:

```
{
    "config": {
        "fail-on-first-error": true
    },
    "pre-commit": {
        "enabled": true,
        "actions": [
            {
              "action": "vendor/bin/php-cs-fixer
                fix --dry-run"
            },
            {
                "action": "vendor/bin/phpstan"
            }
        ]
    }
}
```

Each hook has its own section. Within the pre-commit hook section, enabled can be either true or false—the latter disables the hook but keeps the configuration in the file, which can be handy for debugging purposes. actions contains the actual commands to execute. As you can see, these commands are like the ones you already know from *Chapter 7, Code Quality Tools.*

Every action you want to have executed needs to be written in a separate action section. In the preceding example, we configured PHP-CS-Fixer and PHPStan to be executed on pre-commit.

Since we have additional configuration files for both tools, we do not need the specify any further options, except telling PHP-CS-Fixer to only do a dry run—that is, to only inform us when a code style violation has been found.

In the config section, you can specify further configuration parameters. We want to stop the hook execution immediately after an error occurred, so we set fail-on-first-error to true. Otherwise, CaptainHook would first run all checks, and then tell you the results. This is, of course, just a matter of personal taste.

> **CaptainHook documentation**
>
> We cannot list all features of CaptainHook in this book. However, we encourage you to check out the official documentation at `https://captainhookphp.github.io/captainhook` to learn more about this tool.

As we are done now with the configuration, please store this **JavaScript Object Notation (JSON)** file under the name `captainhook.json` in the project root folder. That is already everything we have to do regarding the configuration.

We just need to install the hooks now—that is, generate hook files in `.git/hooks`. This can be simply done like so:

```
$ vendor/bin/captainhook install -f
```

We use the `-f` option here, which stands for *force*. Without this option, CaptainHook would ask us for every hook separately, if we want to install it. Please note that CaptainHook will install a file for every Git hook it supports, even for those which you did not configure. Those hooks will not do anything, though.

To test the `pre-commit` hook, you can execute this manually without having to commit anything using the following command:

```
$ vendor/bin/captainhook hook:pre-commit
```

Similar commands are available for all the other hooks that CaptainHook supports. If you made changes to the `captainhook.json` file, do not forget to install it by using the `install -f` command again.

To make sure that the hooks get installed in the local development environment, you can add the following code to the `scripts` section of your `composer.json` file:

```
"post-autoload-dump": [
    "if [ -e vendor/bin/captainhook ]; then
      vendor/bin/captainhook install -f -s; fi"
]
```

We use the `post-autoload-dump` event of Composer to run the `install -f` command. The command will be executed every time the Composer autoloader gets refreshed, which will happen every time `composer install` or `composer update` is executed. This way, we make sure that the hooks are installed or updated regularly in the development environments of anybody who works on this project. By using `if [ -e vendor/bin/captainhook ]`, we check if the CaptainHook binary exists and avoid breaking the CI build if it is not installed.

Git hooks in practice

We completed the configuration of the `pre-commit` hook and tested and installed it. Now, we are ready to see it in action: if you do any change in the application code—for example, by adding a blank line somewhere in the `ProductController.php` file—and then try to commit the changes, the `pre-commit` hook should be executed. If the changes violated the *PSR-12* standard, the PHP-CS-Fixer step should fail, as shown in the following screenshot:

```
$ git add src/Controller/ProductController.php
$ git commit -m "Some arbitrary changes"
pre-commit:
 - vendor/bin/php-cs-fixer fix --dry-run                                    : failed

failed to execute: vendor/bin/php-cs-fixer fix --dry-run

   1) src/Controller/ProductController.php

Checked all files in 0.007 seconds, 14.000 MB memory used

Loaded config default from "/home/curtis/dev/github-actions-test/.php-cs-fixer.dist.php".
Using cache file ".php-cs-fixer.cache".

$
```

Figure 11.6: The pre-commit hook fails

Fixing code style issues automatically

Of course, you can remove the `--dry-run` option when executing PHP-CS-Fixer to let it fix issues automatically. In fact, this is a common practice, and we encourage you to try out the same. However, it requires a bit more work, because you must let the user know that their changed files have been fixed and that they need to be re-committed. To keep this example simple, we decided to omit this.

We now know that the `ProductController.php` file has to be fixed. We can let PHP-CS-Fixer do the work, as demonstrated here:

```
$ vendor/bin/php-cs-fixer fix
Loaded config default from "/home/curtis/dev/github-actions-test/.php-cs-fixer.dist.php".
Using cache file ".php-cs-fixer.cache".
   1) src/Controller/ProductController.php

Fixed all files in 0.007 seconds, 14.000 MB memory used
$
```

Figure 11.7: Using PHP-CS-Fixer to automatically fix code style issues

`ProductController.php` has been changed again now, and those additional changes have not yet been staged—that is, they have not yet been added to the commit. The previous changes are still staged, though. The following screenshot shows you what it would look like if you ran `git status` at this point:

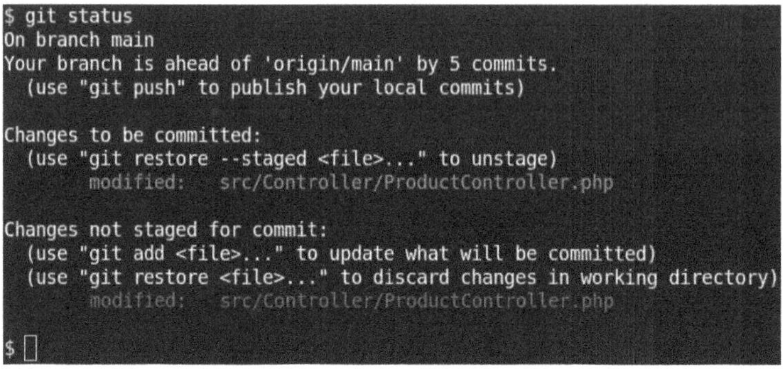

```
$ git status
On branch main
Your branch is ahead of 'origin/main' by 5 commits.
  (use "git push" to publish your local commits)

Changes to be committed:
  (use "git restore --staged <file>..." to unstage)
        modified:   src/Controller/ProductController.php

Changes not staged for commit:
  (use "git add <file>..." to update what will be committed)
  (use "git restore <file>..." to discard changes in working directory)
        modified:   src/Controller/ProductController.php

$
```

Figure 11.8: Unstaged changes

All that needs to be done now is to add the `ProductController.php` file again and run `git commit` again, as demonstrated in the following screenshot:

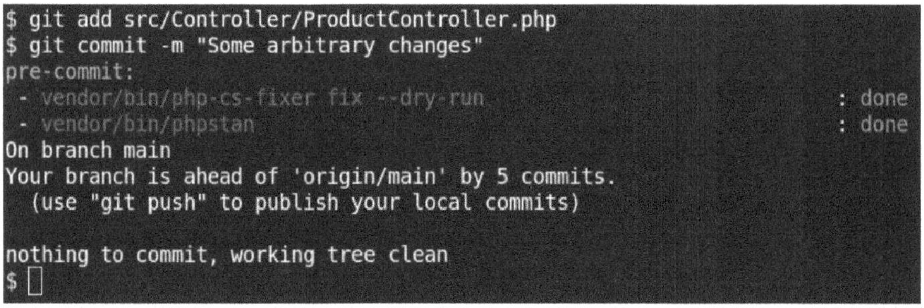

```
$ git add src/Controller/ProductController.php
$ git commit -m "Some arbitrary changes"
pre-commit:
 - vendor/bin/php-cs-fixer fix --dry-run                              : done
 - vendor/bin/phpstan                                                 : done
On branch main
Your branch is ahead of 'origin/main' by 5 commits.
  (use "git push" to publish your local commits)

nothing to commit, working tree clean
$
```

Figure 11.9: pre-commit hook passes

Both steps of the `pre-commit` hook pass now. All you need to do now is to `git push` the committed changes.

Advanced usage

The preceding example was a very basic one. Of course, there is much more that you can do in the local development environment already. You could, for example, add more tools such as the `phpcpd` copy and paste detector or the `phpmd` mess detector, both of which we introduced in *Chapter 7, Code Quality Tools*.

If your tests are not too slow (what exactly that means depends on your and your teammates' patience), you should consider running your tests locally as well. Even if you have slow-running tests, you can separate them into several test suites, and only execute the fast-running tests on `pre-commit`.

You should also consider running code quality checks on modified files only and not the whole project, as we did in our example. CaptainHook provides the useful {$STAGED_FILES} placeholder, which contains all staged files. It is very convenient to use, as we can see here:

```
{
    "pre-commit": {
        "enabled": true,
        "actions": [
            {
                "action": "vendor/bin/php-cs-fixer fix
                    {$STAGED_FILES|of-type:php} --dry-run"
            },
            {

                "action": "vendor/bin/phpstan analyse
                    {$STAGED_FILES|of-type:php}"
            }
        ]
    }
}
```

The preceding example runs the checks only on modified PHP files. This has two main benefits: firstly, it is faster because you do not have to check code that you have not touched. The speedup, of course, depends on the size of your code base.

Secondly, especially if you are working on an existing project and just started introducing these checks, running them on the whole code base is not an option because you would need to fix too many files at once. We will discuss this in more detail in the next section.

Excursion – Adding CI to existing software

If you work in a company, you will not always start *on the green*—that is, build a new project from the ground up. In fact, most likely it will be the opposite: when you join a company, you will be added to a team that has been working on one or more projects for a long time already.

You probably came across the terms *legacy software* or *legacy system* already. In our context, they describe software that has existed for a long time and is still in use in business-critical processes. It does not meet modern development standards anymore, so it cannot be easily updated or changed. Over time, it becomes so brittle and hard to maintain that no developer wants to touch it anymore. What makes it even worse is the fact that because the system grew over a longer time, it has so much functionality that no stakeholder (that is, the users) would like to miss it. So, replacing a legacy system is not that easy.

Not surprisingly, legacy software has a bad connotation, yet probably, all the system needs is some "attention". Think of it like the restoration of an old machine, where old parts get replaced with modern ones, while the outside is left unchanged. Some developers even find it extra challenging to work on such software. It has come a long way, earned its money, and—most likely (at least partly)—pawed the success of the company, so it deserves some respect.

So, if you have to work on such a project, do not give up quickly. With this book, we provided you with the necessary knowledge and tools to start bringing it into shape again—it just will take a bit longer, and you might never reach a perfect level. But being perfect is not necessary anyway.

Step by step

Start with adding integration and E2E tests. Both test types usually require no or just a few changes on the code but will bring great benefit, as they indirectly cover a lot of code without having to write unit tests. Once you've covered the critical paths (that is, the most used workflows) of the application with tests, you can start refactoring the classes and start introducing additional unit tests. The tests will help you to discover bugs and side effects quickly, without having to click through the application repeatedly.

Introducing a code style such as *PSR-12* is, as you know by now, as easy as just running a tool such as `PHP-CS-Fixer` over the entire code base at once. The resulting commit will, of course, be huge, so you want to agree upon a code freeze with any fellow developers before you do it. A code freeze means that everybody commits their changes into the repository so that your refactoring does not cause huge merge conflicts when they check out the changes afterward.

To decide which code to refactor, we intend to use one or more of the many code quality tools you know by now. Going with PHPStan on level 0 is a good choice. You might also want to consider Psalm, as it can also resolve some issues automatically. Depending on the size of the project, the list of errors can be dauntingly long. Making use of the baseline feature, as described in *Chapter 7, Code Quality Tools*, can be a cosmetic help here, but it will only hide and not solve the code issues.

You do not need to rush. If you configure your CI/CD pipeline to only check modified files, you can start improving the code over time, piece by piece. It does leave you with the problem that once you have touched a file, you must refactor it to meet the rules. Especially for old but huge classes, this can be problematic. In *Chapter 7, Code Quality Tools*, however, we explained how you can exclude files or parts of the code from the performed checks. You can even set up a pipeline that allows you to skip the checks upon a certain keyword (for example, `skip ci`) in the commit message. This approach, however, should only be the last resort—otherwise, you will never start refactoring old code. It takes some self-restraint from the developers not to misuse this feature, too.

Over time, the team working on the project will gather new confidence, and with growing test coverage, they will start refactoring more and more code. Make sure to install a local pipeline as well, to keep the waiting times short.

An outlook on CD

Eventually, your CI pipeline will work so well that you can fully trust it. It will prevent shipping broken code into production reliably, and at some point, you find yourself doing fewer and less-manual checks if the deployment went well. At that point, you could think about using CD: this describes the combination of tools and processes to deploy code to any environment automatically.

A usual workflow is that whenever changes get merged into a certain branch (for example, `main` for the production environment), the CI/CD pipeline will be triggered automatically. If the changes pass all checks and tests, the process is trusted so much that the code gets deployed into the desired destination without testing the build result manually anymore.

If you ever had the opportunity to work in such an environment, you surely do not want to miss it. Besides a great CI/CD pipeline and 99% trust in it, it requires some more processes in place to quickly react if a deployment has problems. Even the best tools cannot prevent logical errors or infrastructural issues that will only appear under a greater load.

Whenever there is a problem after deployment, your team should be the first ones to notice! You not only need to fully trust the pipeline but the monitoring and logging setup as well. There are many concepts and tools out there, and we are finally leaving the topic of code quality here, entering the realms of **development-operations** (**DevOps**) and system administration. Nevertheless, we want to give you some short guidance on a few key concepts you might want to delve into, as follows:

- **Monitoring** gathers information about the status of a system. In our context, this is usually information such as **central processing unit** (**CPU**) load, **random-access memory** (**RAM**) usage, or database traffic of all servers or instances. For example, if the CPU load suddenly increases massively, this is an excellent indicator that there is trouble ahead.

- **Logging** helps you organize all log messages your application produces in a single, easily accessible place. It is not helpful if you need to search for any log files on different servers first when the system is in trouble and all alerts are ringing.

- There are multiple **deployment methods** available. Especially when your setup has grown and consists of multiple servers or cloud instances, you can roll out the new code just on a few instances or even a separate deploy environment and monitor the behavior there. If all goes well, you can continue the deploy to the remaining instances. These methods are called **canary**, **rolling**, and **blue/green** deployments. You will find a link with more information on them at the end of this chapter.

- Regardless of how well you monitor your software, if things go wrong (and they will), you need to go back to a previous version of your application. This is called a **rollback**. You should always be prepared to go back to the previous version as fast and easy as possible. This requires you to have the deliverables of several previous versions available. It is a good idea to keep at least 5 or 10 versions because sometimes, it is not clear which version exactly caused the problem.

Surely, CD is beyond the scope of writing clean PHP code. However, we think it is a goal worth aiming for, as it will speed up your development a lot, and introduces you to a variety of fascinating tools and concepts.

Summary

We hope that, after reading this chapter, you are as convinced as we are that CI is extremely helpful and thus a must-have tool in your toolbox. We explained the necessary terms around this topic as well as the different stages of a pipeline, not only in theory but also in practice, by building a simple but working pipeline using GitHub Actions. Finally, we gave you an outlook on CD.

You now have a great foundation of knowledge and tools to write great PHP code. Of course, learning never stops, and there is so much more knowledge out there for you to discover that we could not fit into this book.

If you made developing PHP software your profession, then you usually work in teams of developers. And even if you are maintaining your own open source project, you will interact with others—for example, when they submit changes to your code. CI is an important building block, but not the only thing you need to consider for a successful team setup.

For us, this topic is so important that we dedicated the next two chapters to introducing modern collaboration techniques that will help you to write great PHP code when working in teams. We hope to see you in the next chapter!

Further reading

If you wish to know more, have a look at the following resources:

- Additional information about GitHub Actions:

  - The official *GitHub Actions* documentation with lots of examples: `https://docs.github.com/en/actions`

  - `setup-php` is not only very useful for PHP developers, but also offers a lot of useful information—for example, about the *matrix setup* (how to test code against several PHP versions) or *caching Composer dependencies* to speed up the build: `https://github.com/marketplace/actions/setup-php-action`

- More information about CD and related topics can be found here:

  - A good overview of CD: `https://www.atlassian.com/continuous-delivery`

  - Logging and monitoring explained: `https://www.vaadata.com/blog/logging-monitoring-definitions-and-best-practices/`

- A great introduction to advanced deployment methods: `https://www.techtarget.com/searchitoperations/answer/When-to-use-canary-vs-blue-green-vs-rolling-deployment`

- Tools and links regarding your local pipeline:

  - More insights on Git hooks: `https://git-scm.com/book/en/v2/Customizing-Git-Git-Hooks`

  - GrumPHP is a local CI pipeline "out of the box": `https://github.com/phpro/grumphp`

12

Working in a Team

The main goal of this book is to enable you to write code that can be understood, maintained, and extended by you and others. Most of the time, being a PHP developer means that you do not work alone on a project or a tool. And even if you started writing code alone, chances are high that at some point, another developer will join you – be it on a commercial product, or your open source package where other developers start adding new features or bug fixes.

There will always be multiple ways to carry out a task in software development. This is what makes working in a team more challenging when you want to write *clean code* together. In this chapter, you will find several tips and best practices on how to set up *coding standards* and *coding guidelines*. We will also talk about how *code reviews* will improve the code and ensure the guidelines are kept.

We will also explore the topic of *design patterns* in more detail at the end of this chapter. These patterns can help your team solve typical software development problems because they offer well-tested solutions.

This chapter will include the following sections:

- Coding standards
- Coding guidelines
- Code reviews
- Design patterns

Technical requirements

If you followed along with the previous chapters, you do not require any additional setup.

The code samples for this chapter can be found in our GitHub repository: `https://github.com/PacktPublishing/Clean-Code-in-PHP`.

Coding standards

In the previous chapters, you learned a lot about writing high-quality code. Yet, it is not enough if you do it by yourself only. When you work in a team, you will most likely have the problem that other developers have a different understanding of quality and are on a different skill level than you are.

This is harmful to your code because it might lead to lazy compromises, where the involved parties agree on a way, just to have their peace. Therefore, if you want to work effectively in a team, you want to standardize your work as much as possible.

It makes sense to start with the low-hanging fruit: code formatting. This goes down to the very basics, such as agreeing on how many spaces should be used to indent lines, or where braces should be placed. But why is this even important?

We already shortly addressed this topic in *Chapter 5, Optimizing Your Time and Separating Responsibilities*. However, we want to expand on it at this point. The main advantage of having a common **coding standard** (also called *coding style*) is to reduce *cognitive friction* when reading code.

> **Cognitive friction**
>
> Cognitive friction basically describes the required mental effort for our brain to process information. Imagine, for example, you read a book, where every other paragraph was written in a different font, size, or line spacing. You would still be able to read it, but it would become annoying or tiring soon. The same applies to reading code.

Introducing a coding standard to a project is relatively easy, thanks to the tools we already presented to you earlier in this book. Agreeing with others on a common standard, on the other hand, requires more work. That is why, in this section, we want to show you how to easily align on a common coding standard.

Going with existing standards

Setting up standards together with others can be a long and painful process. However, nowadays, you do not argue about the size of a sheet of paper anymore. In European countries, the *DIN A4* standard is widely accepted, while in other countries, such as the US, you would use the *US Letter Size* without asking why. Most people accept these measures, and following these standards makes life a bit easier – one thing less to care about.

The same applies to coding standards, which define how you format your code. Of course, you could argue for hours with your teammates about whether *tabs* or *spaces* should be used for the indentation. Both sides will come up with valid arguments, and you will never find the right answer, as there simply is no right and wrong here. And once you got the question about indentation sorted, the next topic to discuss could be the placement of brackets. Should they appear in the same line, or in the next?

We do not necessarily need to agree with every detail of a standard, but undoubtedly, it saves time and nerves to use existing norms. In the PHP ecosystem, there are *Coding Standards* that already exist that you could utilize. A huge additional benefit of doing so is that the code sniffers have built-in rule sets for these standards. In the next section, we will talk about probably the best-known *Coding Standard* for PHP.

PHP-FIG and PSR

PHP itself has no official *Coding Standard*. Historically, each major PHP framework that existed, or still exists today, introduced some sort of standards because the developers quickly realized that using them has its benefits.

However, since every project used its own standards, the PHP world ended up with a mixture of different formatting standards. Back in 2009, when the **PHP-FIG (PHP Framework Interoperability Group (PHP-FIG)** was formed, which consisted of members from all the important PHP projects and frameworks of that time, they wanted to solve exactly these kinds of problems.

At that time, Composer was becoming more and more important, and packages were introduced that could easily be used across different frameworks. To keep the code somewhat consistent, a mutual way to write code was agreed upon: the **PHP Standard Recommendations (PSRs)** were born.

To make the autoloader of Composer work, it was necessary to agree on how to name classes and directories. This was done with the very first standard recommendation, *PSR-0* (yes, nerds start counting at 0), which was eventually replaced by *PSR-4*.

The first *Coding Standard* recommendation was introduced with *PSR-1* and *PSR-2*. *PSR-2* was later replaced by *PSR-12*, which contained rules for language features of newer PHP versions.

Although *PSR-12* addresses the code style, it does not cover naming conventions or how to structure the code. This is often still predefined by the framework you use. The *Symfony* framework, for example, has its own set of *Coding Standards* that are based on the aforementioned *PSR-4* and *PSR-12*, but add further guidelines, for example, conventions on naming or documentation. Even if you do not use a framework at all and just pick single components to build an application, you could consider using these guidelines, which you will find on the *Symfony* website: `https://symfony.com/doc/current/contributing/code/standards.html`.

> **PER coding style**
>
> *PSR-12* was released in 2019 and thus does not cover the latest PHP features anymore. Therefore, at the time of writing this book, PHP-FIG released the *PER Coding Style 1.0.0* (**PER** is short for **PHP Extended Recommendation**). It is based on *PSR-12* and contains some additions to it. In the future, the PHP-FIG no longer plans to release any new *Coding Standards* related to PSRs, but new versions of this PER, if it is required. It is very likely that the **code quality** tools we featured in this book will pick up the new PER soon. You will find more information about it here: `https://www.php-fig.org/per/coding-style`.

Over time, the PHP-FIG has introduced over a dozen recommendations, and more are in the making. They cover topics such as how to integrate logging, caching, and HTTP clients, to just name a few. You will find a complete list on the official website: `https://www.php-fig.org`.

Problems with PHP-FIG and PSR

The PHP-FIG should not be considered the official PHP authority, and neither should any PSR be taken as indisputable. In fact, many important frameworks such as *Symfony* or *Laravel* are not part of the PHP-FIG anymore, since the recommended standards have interfered too much with their internals. Looking at all the PSRs that are available today, you could even regard them as their own meta-framework. This is not to diminish the relevance of many recommendations though – we just want you to not blindly accept them as granted.

Enforcing coding standards in your IDE

There are several ways to enforce coding standards. In the previous chapter, *Chapter 11, Continuous Integration*, we explained how to make sure that no wrongly formatted code can spoil the code base. This worked fine, yet it requires an additional step, even if we let our tools fix the code formatting automatically because we need to commit those changed files again. So, would it not actually be useful if our code editor, or IDE, would help us with formatting the code while we write it?

Modern code editors usually have built-in functionality that assists you with adhering to your preferred coding standards, if you configure them. If not built-in, this functionality can at least be provided with plugins.

There are two basic ways your editor could support you:

- **Highlighting the coding standard violations**: The IDE marks those parts of the source code that need to be corrected. It will not change the code actively though.

- **Reformatting the code**: Either the IDE or an additional plugin takes care of formatting the code, for example, by running a code style fixer such as `PHP_CS_Fixer`. This can be done upon manual request, or every time a file is saved.

Reformatting the code on file save is a very convenient way to ensure that your code meets the coding standards. How to set this up depends on which IDE you are using, so we will not elaborate on this further in this book.

We would still recommend using **Git hooks** and **continuous integration** as the second layer of checks to make sure no badly formatted code gets pushed to the project repository. You can never be sure whether a team member accidentally or willingly disabled the automated reformatting or did not care about the highlighted parts of the code.

Coding Standards are all about how to format code consistently. But that is not all you should agree on when working in a team – in the next section, we will show you what other aspects are worth agreeing upon.

Coding guidelines

In the previous section, we talked about why you should introduce *Coding Standards*. Once this is accomplished, you should consider setting up **coding guidelines**. Both topics sound very familiar, and indeed, they are. Yet while *Coding Standards* usually focus on how to format code, coding guidelines define how to write code. This, of course, includes defining which *Coding Standard* to use, but covers a lot more, as you will learn in this section.

What does *how to write code* exactly mean? Usually, there is more than one way to achieve things when writing software. Take the widely known **model-view-controller** (**MVC**) pattern, for example. It is used to divide the application logic into three types of interconnected elements – the models, the views, and the controllers. It does not explicitly define where to place the **business logic**, though. Should it be located inside the controllers, or rather inside the models?

There is no clear right or wrong answer to this question. Our recommendation, however, would be the *fat models, skinny controllers* approach: business logic should *not* be written within the controllers, as they are the binding element between the views and your problem-specific code. Also, the controllers usually contain a lot of framework-specific code, and it is good practice to keep that out of your business logic as much as possible.

Regardless of our recommendation, it should be defined in the coding guidelines of your project how you think your team should handle this question. Otherwise, you will most likely end up having both approaches in your code base.

Usually, coding guidelines cover questions such as how to name methods, functions, and properties. As you might know from the famous quote "*There are only two hard things in computer science: cache invalidation and naming things,*" finding the right names is indeed not a trivial problem. So, having conventions on this topic at least reduces the time of developers trying to come up with a proper name. Furthermore, the same as *Coding Standards*, they help reduce cognitive friction.

Coding guidelines help the lesser-experienced developers in your team, or those who just started, to have a solution at hand that they otherwise needed to search for in the code or on the internet. It also helps write maintainable code by avoiding bad practices, as we already discussed in *Chapter 3, Code, Don't Do Stunts*. To help you get started with setting up the first set of rules, we will give you some examples in the next section.

Examples of coding guidelines

Starting with a blank sheet of (virtual) paper is hard, so in this section, we collected a list of real-world examples of what could be part of your coding guidelines. Please note that this collection of rules, although they are based on best practices, is not meant to be perfect or the only truth. We rather want to give you a good starting point for discussions and examples on what topics should be clarified using coding guidelines.

Naming conventions

By using **naming conventions**, we make sure that certain elements of our code are being named in a uniform and comprehensible way. This reduces cognitive friction and makes the onboarding of new team members easier.

Services, repositories, and models

Written in *UpperCamelCase*. Use the type as the suffix.

Here are some examples:

- `UserService`
- `ProductRepository`
- `OrderModel`

Events

Written in *UpperCamelCase*. Use the correct tense to indicate whether the event is fired before or after the actual event.

Here are some examples:

- `DeletingUser` is the event before the deletion
- `DeleteUser` is the actual event
- `UserDeleted` is the event after the deletion

Properties, variables, and methods

Written in *lowerCamelCase*.

Here are some examples:

- `$someProperty`
- `$longerVariableName`
- `$myMethod`

Tests

Written in *lowerCamelCase*. Use the word *test* as a prefix.

Here are some examples:

- `testClassCanDoSomething()`

Traits

Written in *UpperCamelCase*. Use the adjective to describe what the trait is used for.

Here are some examples:

- `Loggable`
- `Injectable`

Interfaces

Written in *UpperCamelCase*. Use the word *Interface* as the suffix.

Herer are some examples:

- `WriterInterface`
- `LoggerInterface`

General PHP conventions

Even if you already use *Coding Standards* such as *PSR-12*, there are certain aspects that they do not cover. We will pick up some of them in this section.

Comments and DocBlocks

Avoid comments if possible, as they tend to get outdated and thus confuse more than they help. Only keep comments that cannot be replaced by self-explanatory names or by simplifying code, so it is easier to understand and does not require the comment anymore.

Only add **DocBlocks** if they add information, such as annotations for the code quality tools. Particularly since PHP 8, most DocBlocks can be replaced by type hints, which all modern IDEs will understand. If you use type hints, most DocBlocks can be removed:

```
// Redundant DocBlock

/**
 * @param int $property
 * @return void
```

```
    */
    public function setProperty(int $property): void {
        // ...
    }
```

Often, DocBlocks are automatically generated by the IDE. If they are not updated, they are at best useless, or can even be plainly wrong:

```
// Useless DocBlock

/**
 * @param $property
 */
public function setProperty(int $property): void {
    // ...
}


// Wrong DocBlock

/**
 * @param string $property
 */
public function setProperty(int $property): void {
    // ...
}
```

DocBlocks should still be used for information that cannot be provided by PHP language features until now, such as specifying the content of an array, or marking a function as deprecated:

```
// Useful DocBlock

/**
 * @return string[]
 */
public function getList(): array {
    return [
        'foo',
        'bar',
```

```
        ];
    }


    /**
     * @deprecated use function fooBar() instead
     */
    public function foo(): bool {
        // ...
    }
```

> **About DocBlocks**
>
> DocBlocks were introduced to, among other things, partially compensate for the shortcomings of weak typing in earlier versions of PHP. The de facto standard was introduced by the phpDocumentor project (https://www.phpdoc.org/) and as such is supported by many tools such as IDEs and static code analyzers. Using strict typing, it is often not necessary to use DocBlocks anymore though, unless you want to use phpDocumentor in your project.

Ternary operators

Every part should be written in a single line to increase readability. Exceptions can be made for very short statements:

```
// Example for short statement
$isFoo ? 'foo' : 'bar';


// Usual notation
$isLongerVariable
    ? 'longerFoo'
    : 'longerBar';
```

Do not use nested ternary operators, as they are hard to read and debug:

```
// Example for nested operators
$number > 0 ? 'Positive' : ($number < 0 ? 'Negative' :
'Zero');
```

Constructor

Use **constructor property promotion** for shorter classes, if working with PHP 8+. Keep the trailing comma after the last property, as this will make it easier to add or comment out lines:

```
// Before PHP 8+
class ExampleDTO
{
    public string $name;
    public function __construct(
        string $name
    ) {
        $this->name = $name;
    }
}

// Since PHP 8+
class ExampleDTO
{
    public function __construct(
        public string $name,
    ) {}
}
```

Arrays

Always use the short array notation and keep the comma after the last entry (see the previous section, *Constructor*, for an explanation):

```
// Old notation
$myArray = array(
    'first entry',
    'second entry'
);

// Short array notation
$myArray = [
    'first entry',
```

```
    'second entry',
];
```

Control structures

Always use brackets, even for one-liners. This reduces cognitive friction and makes it easier to add more lines of code later:

```
// Bad
if ($statement === true)
    do_something();

// Good
if ($statement === true) {
    do_something();
}
```

Avoid else statements and return early, as this is easier to read and reduces the complexity of your code:

```
// Bad
if ($statement) {
    // Statement was successful
    return;
} else {
    // Statement was not successful
    return;
}

// Good
if (!$statement) {
    // Statement was not successful
    return;
}

// Statement was successful
return;
```

Exception handling

Empty `catch` blocks should be avoided, as they silently swallow error messages and thus can make it difficult to find bugs. Instead, log the error message or at least write a comment that explains why the exception can be ignored:

```
// Bad
try {
    $this->someUnstableCode();
} catch (Exception $exception) {}

// Good
try {
    someUnstableCode();
} catch (Exception $exception) {
    $this->logError($exception->getMessage());
}
```

Architectural patterns

Coding guidelines are not limited to how to format code or name elements. They can also help you to control how the code is written in an architectural sense.

Fat models, skinny controllers

If the MVC pattern is used, the business logic should be located inside **models** or similar classes, such as **services** or **repositories**. **Controllers** should contain as little code as possible as is required to receive or transfer data between the **views** and the **models**.

> **Framework-agnostic code**
>
> In the context of the *fat models, skinny controllers* approach, you will probably come across the term *framework-agnostic business logic*. It means that the code that contains your business rules should use as few features of the underlying framework as possible. This makes framework updates or even migrations to other frameworks much easier.

Single responsibility principle

Classes and methods should only have one responsibility. See *Chapter 2, Who Gets to Decide What "Good Practices" Are?*, for more information about this principle.

Framework guidelines

In this book, we want to focus on writing clean code in PHP. Often, though, you will be working with frameworks, and although it is essential to include them in the guidelines as well, we do not want to go into much more detail here.

However, next, you will find a list of questions that should give you a good idea about which framework-related topics to include in your guidelines:

- How to access the database
- How to configure routes
- How to register new services
- How is authentication handled within your project?
- How should errors or other debug information be logged?
- How to create and organize view files
- How to handle translations

Since the answers to these questions highly depend on the used framework, we cannot give you recommendations here. You will need to set the guidelines up together with your team. In the next section, we will give you some ideas on how to do that.

Setting up guidelines

The process of setting up coding guidelines takes time and often requires several workshops in which the rules are discussed. This requires moderation, for example, by a technical lead; otherwise, you might get stuck in endless discussions.

Do not worry if you cannot immediately reach an agreement on all topics though. Remind yourself that the people in your team have different backgrounds, experiences, and skill levels – and no one will directly ditch their personal ways of coding just because there are suddenly rules that they do not understand or accept.

Make sure to set up a process that checks from time to time whether the guidelines need to be updated. Maybe some rules get outdated over time, or new language features must be included. An action point in a regularly occurring team meeting would be a good opportunity for this.

The guidelines should be easily accessible in written form, such as in a wiki or the company's internal knowledge base, which should be able to track the version history. Every team member should be able to write comments on it so that questions or issues can be handled as soon as they appear. Lastly, all team members should be automatically informed about new changes.

Once your team agrees on a set of rules, make sure to utilize the code quality tools you learned about in earlier chapters to automatically check whether the rules are respected. You can, for example, use *PHPStan* to detect empty `catch` blocks, or *PHPMD* to enforce `if` without using `else`.

How can we ensure that our coding guidelines are applied? Obviously, we should use our code quality tools wherever possible. But what if these tools do not include the rules we would like to enforce? With a bit of internet research, you might be able to find a third-party implementation for them. Or, if you cannot find anything, you could even write custom rules yourself, since all static code analyzers are extendable.

For rules that are too complicated to check automatically, we have to manually check whether they are used correctly. This can happen in code reviews, and we think they are so important that they deserve their own section in this chapter.

Setting up coding guidelines alone will just be a waste of time if you do not make sure that they are kept. We can automate checking all the coding style-related rules and also a fair number of coding guidelines. But at the moment, for those rules that are addressing the framework guidelines or architectural aspects, automation is no longer possible, and we humans must jump in, taking over the checks. At this point, code reviews come into play. Let us have a closer look in the next section.

Code reviews

The process of manually checking the code of other developers is called a **code review**. This includes all changes, that is, not only new functionality but also bug fixes or even simple configuration changes.

A review is always done by at least one fellow developer, and it usually happens in the context of a **pull request**, shortly before the code of a feature or bug fix branch gets merged into the `main` branch; only if the reviewer approves the changes will they become part of the actual application.

In this section, we will discuss what you should look for in code reviews, why they are so important, and how they should be done to make them a successful tool in your toolkit.

Why you should do code reviews

It might sound a bit obvious because that is what this whole book is about. Yet, it cannot be stressed enough – code reviews will improve the quality of your code. Let us examine more closely why:

- **Easy to introduce**: Introducing code reviews usually comes with no additional costs (except for the required time). All major Git repository services such as **Bitbucket**, **GitLab**, or **GitHub** have a built-in review functionality that you can use immediately.

- **Quick impact**: Code reviews are not only easy to introduce but they will show their usefulness very soon after they have been introduced.

- **Knowledge sharing**: Because code reviews often lead to discussions between the developers, they are a great tool to spread knowledge about best practices in the team. Of course, junior developers especially will massively benefit from the mentoring, but also the most seasoned developers will learn something new from time to time.

- **Constant improvement**: The regular discussions will result in improved coding guidelines, as they are constantly challenged and updated, if necessary.

- **Avoid problems early**: Code reviews take place very early in the process (see *Chapter 11, Continuous Integration*), so chances are good that bugs, security issues, or architectural problems are found before they even reach the test environment.

If you are not yet convinced of the benefits of code reviews, check out the next section, in which we will talk more about what code reviews should cover – and what not.

What code reviews should cover

What aspects should we check when doing code reviews?

- **Code design**: Is the code well designed and consistent with the rest of the application? Does it follow general best practices, such as reusability, design patterns (see the next section), or **SOLID** design principles (see *Chapter 2, Who Gets to Decide What "Good Practices" Are?*)?

- **Functionality**: Does the code do what it should or does it have any side effects?

- **Readability**: Is the code easy to understand or too complex? Are the comments necessary? Could the readability be improved by renaming a function or a variable, or by extracting code into a function with a meaningful name?

- **Security**: Does the code introduce potential attack vectors? Is all output escaped to prevent XSS attacks? Are database inputs sanitized to avoid SQL injections?

- **Test coverage**: Is the new code covered with automated tests? Do they test the right things? Are more test cases needed?

- **Coding standards and guidelines**: Does the code follow the *Coding Standards* and coding guidelines the team agreed upon?

Your team should also consider whether testing the code in their local development environment should be part of the review process or not. There is no clear recommendation on this though.

Best practices for code reviews

Although code reviews have many benefits and can be implemented fairly easily, there are a few pitfalls that you should be aware of, and established best practices that will make the reviews even more successful.

Who should review the code?

First and foremost, who should ideally be doing the *code reviews*? Of course, this also depends on your setup. If you work in a team together with another PHP developer, then this should surely be the first person to ask. This way, you build up shared domain knowledge; although your colleague has not worked on your ticket directly, they at least get an idea of what you have worked on.

Yet, reaching out to members of other teams (if there are any) from time to time avoids being stuck in a bubble and fosters knowledge sharing. If you are unsure about certain topics, ask the domain experts for their assistance. Often, this includes performance, architecture, or security-related changes.

Automatize

One thing that code reviews should not cover is whether the *Coding Standards* are kept. In *Chapter 7, Code Quality Tools* we introduced the necessary tools to do this automatically, and in *Chapter 11, Continuous Integration*, we integrated them into a **CI pipeline**.

Make sure that only those pull requests get reviewed where all the checks (such as *code sniffers, code analyzers, and automated tests*) have passed. Otherwise, you will spend a lot of time on topics that should not even be discussed.

Avoid long code reviews

How many lines should the code change that needs to be reviewed have? Studies suggest that 200 to 400 lines should be the maximum, as the concentration of the reviewer decreases over time. So, try to keep the individual changes relatively small. It is also much more likely for the reviewer to find time to review smaller changes, as for a long tapestry of diffs.

Code reviews, even smaller ones, will require time in which the reviewer will not be able to write code. But how much time should be spent? Again, this depends on your setup. A good ballpark number is a maximum of 60 minutes to avoid the fatigue of the reviewer. Allow enough space for the reviewer to review the code line by line. Reviews should be accepted as part of your daily work or they will quickly become a burden, so nobody should rush through them.

Stay human

How to formulate feedback is crucial to make reviews successful. Watch your tone and try to avoid accusations such as "*This is wrong!*" or an absolute no-go, "*This is stupid.*" Developers, especially the lesser experienced ones, should not be anxious to let their code be reviewed.

Remember that a human being will read your comments. It often works well if you write them from the "I" perspective, for example, "*I do not understand this line, can you please explain?*" or "*I think we could also do it like this...*"

Do not forget to use the reviews to give praise for parts that are well done. A quick "*Great idea!*" or "*I really like your approach*" or "*Thanks for the code cleanup*" shows your appreciation of the other developer's work and increases their motivation.

Usually, code reviews are done by writing comments on the Git platform you use. But of course, you can also do them face to face. Some developers appreciate direct feedback more than just comments because written text lacks a lot of meta information, such as the tone of voice or the facial expression.

Don't overdo it, but don't be careless either

Remember the **Pareto principle** and do not overdo things. Maybe there are still small parts in the code that you would change but that are not explicitly wrong, as they adhere to all the team standards. Programming is still a matter of personal style, and having endless discussions in a code review will lead to frustration without further benefit.

Do not accept changes that degrade the overall system health though. If you are convinced that a change is harmful or violates the coding guidelines, you must not approve the changes. If in doubt, get another developer involved.

Embrace changes

Lastly, if you feel an issue that you discussed in a review should be part of the guidelines, note it down and address it in the next team meeting, without mentioning the other developer directly. Maybe you were right and the guidelines will be amended to avoid the issue in the future.

But you could also be wrong, and the rest of the team does not see it as a problem. If you cannot come up with convincing arguments and examples, you have to accept those decisions as well.

Ensuring code reviews are done

During stressful daily work, with high-priority bug fixes and tight deadlines, it is easy to forget about doing code reviews. Fortunately, all the Git service providers offer functionality to assist you here:

- **Make reviews mandatory**: Most Git repository services can be configured to allow changes only to be merged into the `main` branch if they got at least one approval. You should definitely enable this feature.

- **Rotate reviews**: If your team is larger, try to request the review not always from the same person. Some tools even allow selecting a reviewer randomly for you.

- **Use a checklist**: Checklists have proven to be useful, so you should use them too. Set up a checklist for all the aspects you need to look for in code reviews. In the next section, we will show you how to make sure it gets used.

Definition of done

If you work using **agile methodologies**, you probably heard the term *definition of done* already. Here, the team agrees on a list of actions that should be done before a task is completed.

A typical *definition of done* contains checks such as whether the test has been written or the documentation updated. You can utilize this for the code reviews as well.

Again, our Git tools help us by providing templates for pull requests (also called **merge requests**). These are texts that will be used to automatically prefill the description of the pull request.

How that works depends on the software you use, so we cannot give you exact instructions here. The following text, however, shows you an example of what it could look like:

```
# Definition of Done


## Reviewer
[ ] Code changes reviewed
    1. Coding Guidelines kept
    2. Functionality considered
    3. Code is well-designed
    4. Readability and Complexity considered
    5. No Security issues found
    6. Coding standard and guidelines kept
[ ] Change tested manually
## Developer
[ ] Acceptance Criteria met
[ ] Automated Tests written or updated
[ ] Documentation written or updated
```

What gets included in the checklist is up to you and your team. If used as a template, these items will always appear in the pull request description by default. It is meant to be used by both the reviewer and the developer to not forget what needs to be done before approving the pull request and merging it into the `main` branch.

Some tools, such as GitHub, use a **Markdown**-style markup language for these templates. They will display the checkboxes (the two square brackets before every item) as clickable checkboxes in the browser and keep track of whether they were clicked or not. Voilà! Without much work, you have set up an easy-to-use and helpful checklist!

Code reviews conclusion

We hope this section gave you good insights into how beneficial code reviews can be for your team and yourself. Since they can be introduced effortlessly, it is worth trying them out. The best practices in this section will help you avoid some of the problems that code reviews could have.

But, as always, they also have some downsides: reviews take a lot of time and they can lead to conflicts between team members. We are convinced that the time spent pays off well though because the positive aspects outweigh the negative ones by far. The conflicts would most likely happen anyway, and the reviews are just the gauge to vent off steam. This cannot be fully avoided if you work in a team but should be addressed early with your manager. It is their job to deal with these kinds of problems.

In the last part of this chapter, we will look at design patterns in more detail. They can act as guidelines on how to solve general problems in software development.

Design patterns

Design patterns are commonly used solutions to problems that occur regularly in software development. As a developer, you will sooner or later come across this term, if you have not done so already – and not without a reason, as these patterns are based on best practices and have proven their usefulness.

In this section, we will tell you more about the different types of design patterns and why they are so important that they became part of this book. Furthermore, we will introduce you to some common design patterns that are widely used in PHP.

Understanding design patterns

Let us have a closer look at design patterns now. They can be considered templates to solve particular problems and are named according to the solution they provide. For example, in this chapter, you will learn about the **Observer** pattern, which can help you to implement a way to observe changes in objects. This is very useful when you write code, but also when you design software with other developers. It is much easier to use a short name to name a concept rather than having to explain it every time.

Do not mistake design patterns with algorithms though. Algorithms define clear steps that need to be followed to solve a problem, while design patterns describe how to implement the solution on a higher level. They are not bound to any programming language.

You also cannot add design patterns to your code like you would add a Composer package, for example. You have to implement the pattern on your own, and you have certain degrees of freedom in how you do that.

However, design patterns are not the single solution to every problem, nor do they claim to offer the most efficient solutions. Always take these patterns with a grain of salt – often, developers want to implement a certain pattern just because they know it. Or, as the saying goes: *"If all you have is a hammer, everything looks like a nail."*

Usually, design patterns are divided into three categories:

- **Creational patterns** deal with how to efficiently create objects and at the same time offer solutions to reduce code duplication

- **Structural patterns** help you organize relationships between entities (i.e., classes and objects) in flexible and efficient structures

- **Behavioral patterns** arrange communication between entities while maintaining a high degree of flexibility

In the following pages, we will have a look at some example implementations to explain the idea behind *design patterns*.

Common design patterns in PHP

We now want to introduce some of the most widely used design patterns in the PHP world. We chose one pattern each from the three categories **Creational**, **Structural**, and **Behavioral**, which we discussed in the previous section.

Factory Method

Imagine the following problem: you need to write an application that should be able to write data into files using different formats. In our example, we want to support **CSV** and **JSON**, but potentially, other formats in the future as well. Before the data is written, we would like to apply some filtering, which should always happen, regardless of which output format is chosen.

An applicable pattern to solve this problem would be the **Factory Method**. It is a Creational pattern, as it deals with the creation of objects.

The main idea of this pattern is that subclasses can implement different ways to achieve the goal. It is important to note that we do not use the new operator in the parent class to instantiate any subclasses, as you can see in the following class:

```
abstract class AbstractWriter
{
    public function write(array $data): void
    {
        $encoder = $this->createEncoder();

        // Apply some filtering which should always happen,
        // regardless of the output format.
        array_walk(
            $data,
```

```
            function (&$value) {
                $value = str_replace('data', '', $value);
            }
        );


        // For demonstration purposes, we echo the result
        // here, instead of writing it into a file
        echo $encoder->encode($data);
    }


    abstract protected function createEncoder(): Encoder;
}
```

Note the `createEncoder` method – this is the factory method that gave the pattern the name, since it acts, in a sense, as a factory for new instances. It is defined as an abstract function, so it needs to be implemented by one or more subclasses.

To be flexible enough for future formats, we intend to use separate `Encoder` classes for each format. But first, we define an interface for these classes so that they are easily exchangeable:

```
interface Encoder
{
    public function encode(array $data): string;
}
```

Then, we create an `Encoder` class for each format that implements the `Encoder` interface; first, we create `JsonEncoder`:

```
class JsonEncoder implements Encoder
{
    public function encode(array $data): string
    {
        // the actual encoding happens here
        // ...

        return $encodedString;
    }
}
```

Then we create CsvEncoder:

```
class CsvEncoder implements Encoder
{
    public function encode(array $data): string
    {
        // the actual encoding happens here
        // ...

        return $encodedString;
    }
}
```

Now, we need to create one subclass of the AbstractWriter class for each format we want to support. In our case, that is CsvWriter first:

```
class CsvWriter extends AbstractWriter
{
    public function createEncoder(): Encoder
    {
        $encoder = new CsvEncoder();
        // here, more configuration work would take place
        // e.g. setting the delimiter
        return $encoder;
    }
}
```

And second, it is JsonWriter:

```
class JsonWriter extends AbstractWriter
{
    public function createEncoder(): Encoder
    {
        return new JsonEncoder();
    }
}
```

Please note that both subclasses only overwrite the Factory Method `createEncoder`. Also, the new operators occur only in the subclasses. The `write` method remains unchanged, as it gets inherited from `AbstractWriter`.

Finally, let us put this all together in an example script:

```
function factoryMethodExample(AbstractWriter $writer)
{
    $exampleData = [
        'set1' => ['data1', 'data2'],
        'set2' => ['data3', 'data4'],
    ];

    $writer->write($exampleData);
}

echo "Output using the CsvWriter: ";

factoryMethodExample(new CsvWriter());

echo "Output using the JsonWriter: ";

factoryMethodExample(new JsonWriter());
```

The `factoryMethodExample` function first receives `CsvWriter` and, for the second run, `JsonWriter` as parameters. The output will look like this:

```
Output using the CsvWriter:
3,4
1,2

Output using the JsonWriter:
[["3","4"],["1","2"]]
```

The Factory Method pattern enables us to move the instantiation of the `Encoder` class away from the `AbstractWriter` parent class into the subclasses. By doing this, we avoid tight coupling between `Writer` and `Encoder`, gaining much more flexibility. As a downside, the code becomes more complex, as we have to introduce interfaces and subclasses to implement this pattern.

Dependency injection

The next pattern we want to introduce is a Structural pattern called **Dependency Injection** (**DI**). It helps us to implement a loosely coupled architecture by inserting dependencies into classes already at construction time, instead of instantiating them inside the class.

The following code shows you how a dependency, in this example, a classic **Logger**, gets instantiated within the constructor:

```
class InstantiationExample
{
    private Logger $logger;

    public function __construct()
    {
        $this->logger = new FileLogger();
    }
}
```

The code itself works perfectly fine, yet the problems start when you want to replace `FileLogger` with a different class. Although we already use the `Logger` interface for the `$logger` property, which theoretically makes it easy to exchange it with another implementation, we have hardcoded `FileLogger` in the constructor. Now, imagine you used that logger in almost every class; replacing it with a different `Logger` implementation causes some effort, as you would have to touch every single file that uses it.

Not being able to replace `FileLogger` also makes writing **unit tests** for the class more difficult. You cannot replace it with a mock, but you also do not want to write information to your actual logs during test runs. If you want to test that the logging works correctly, you have to build quite some workarounds into your code that is also used in production.

DI forces us to think about which, and how many, dependencies should be used in a class. It is considered a **code smell** (i.e., an indicator for badly structured code) when the constructor takes considerably more than three or four dependencies as a parameter because it indicates that the class violates the **single responsibility principle** (the "S" in **SOLID**). This is also known as **scope creep**: the scope of a class slowly but steadily gets bigger over time.

Let us now see how DI would solve the previously mentioned problems:

```
class ConstructorInjection
{
    private Logger $logger;
```

```
    public function __construct(Logger $logger)
    {
        $this->logger = $logger;
    }
}
```

> **Constructor property promotion**
>
> Please note that we did not use constructor property promotion purposely here for better visualization.

The difference to the previous code does not seem to be that big. All we did was pass over the `Logger` instance as a parameter to the constructor instead of instantiating it there directly. The benefit is huge though: we can now change the instance that gets injected (if it implements the `Logger` interface) without touching the actual class.

Imagine you no longer want the class to log into the filesystem, but rather into a **log management system** such as **Graylog**, which manages all the logs from your different applications in one place. All you need to do is to create `GraylogLogger`, which implements the `Logger` interface as well but writes the logs to this system instead of into files. Then, you simply inject `GraylogLogger` instead of `FileLogger` into all classes that should use it – congratulations, you just changed the way your applications log information without touching the actual classes.

Likewise, we can easily exchange a dependency with a mock object in our unit tests. This is a massive improvement regarding testability.

The instantiation of `Logger`, however, no matter which implementation you chose, still has to happen somewhere else. We just moved it out of the `InjectionExample` class. The dependency gets injected when the class gets instantiated:

```
$constructorInjection = new ConstructorInjection(
    new FileLogger()
);
```

Usually, you would find this kind of instantiation within a `Factory` class. This is a class that implements, for example, the **Simple Factory** pattern and whose only job is to create instances of a certain class, with all the necessary dependencies.

> **Simple Factory pattern**
>
> We do not discuss this pattern in more detail in this book because it is, well, really simple. You can find more information about it here: https://designpatternsphp.readthedocs.io/en/latest/Creational/SimpleFactory/README.html.

The injection does not necessarily need to happen through the constructor. Another possible approach is the so-called **setter injection**:

```
class SetterInjection
{
    private Logger $logger;

    public function __construct()
    {
        // ....
    }

    public function setLogger(Logger $logger): void
    {
        $this->logger = $logger;
    }
}
```

The injection of the dependency would then happen using the setLogger method. The same as for the **constructor injection**, this would most likely happen within the Factory class.

The following is an example of what such a **factory** could look like:

```
class SetterInjectionFactory
{
    public function createInstance(): SetterInjection
    {
        $setterInjection = new SetterInjection();
        $setterInjection->setLogger(new FileLogger());

        return $setterInjection;
    }
}
```

Dependency injection container

You probably already wondered how to manage all the factories that are necessary, especially in a larger project. For this, the **DI container** has been invented. It is not part of the DI pattern, yet closely related, so we want to introduce it here.

The DI container acts as central storage for all objects that are brought into their target classes using DI patterns. It also contains all the necessary information to instantiate the objects.

It also can store created instances so it does not have to instantiate them twice. For example, you would not create a `FileLogger` instance for each class that uses it, as this would end up in plenty of identical instances. You rather want to create it once and then pass it over by reference to its destination classes.

DI container

Showing all the functionality of a modern DI container would exceed this book. If you are interested in learning more about this concept, we recommend you to check out the `phpleague/container` package: `https://container.thephpleague.com`. It is small yet feature-rich and has great documentation that can introduce you to more exciting concepts such as service providers or inflectors.

The concept of the DI container has been adopted in all major PHP frameworks nowadays, so you have most likely used such a container already. You probably did not notice it though, because it is usually hidden deep in the back of your application and is sometimes also referred to as a **service container**.

PSR-11 – Container interface

The DI container is so important to the PHP ecosystem that it got its own PSR: `https://www.php-fig.org/psr/psr-11`.

Observer

The last pattern we want to introduce in this book is the **Observer** pattern. As a *Behavioral pattern*, its main purpose is to allow efficient communication between objects. A common task to implement is to trigger a certain action on one object when the state of another object changes. A state change could be something as simple as a value change of a class property.

Let us start with another example: you have to send out an email to the sales team every time a customer cancels their subscription so that they get informed and can do countermeasures to keep the customer.

How do you best do that? You could, for example, set up a recurring job that checks within a certain time interval (e.g., every 5 minutes) whether there had been any cancellations since the check. This would work, but depending on the size of your customer base, the job would probably not return any results most of the time. If the interval between two checks, on the other hand, is too long, you might lose valuable time until the next check.

Now, sales might not be the most time-critical thing in the world (salespeople usually disagree here), but you surely get the idea. Wouldn't it be great if we could just send out the email as soon as the customer cancels the subscription? So, instead of regularly checking for changes, we only get informed at the moment when the change happens?

The code could look like this simplified example:

```
class CustomerAccount
{
    public function __construct(
        private MailService $mailService
    ) {}

    public function cancelSubscription(): void
    {
        // Required code for the actual cancellation
        // ...

        $this->mailService->sendEmail(
            'sales@example.com',
            'Account xy has cancelled the subscription'
        );
    }
}
```

> **Simplified example**
> The example has been simplified. You should, for example, not hardcode the email address.

That approach would surely work, but it has a drawback: the call of `MailService` is directly coded into the class and hence is tightly coupled to it. And now, the `CustomerAccount` class has to care about another dependency, which increases the maintenance effort, as the tests have to be extended, for example. If we later do not want to send this email anymore or even send an additional email to another department, `CustomerAccount` has to be changed again.

Using a loosely coupled approach, the `CustomerAccount` object would only store a list of other objects that it should notify in case of a change. The list is not hardcoded, and the objects that need to get notified have to be attached to that list during the bootstrap phase.

The object that we want to be observed (in the preceding example, `CustomerAccount`), is called the **subject**. The subject is responsible for informing the **observers**. No code change would be necessary on the subject to add or remove observers, so this approach is very flexible.

The following code shows an example of how the `CustomerAccount` class could implement the **Observer** pattern:

```php
use SplSubject;
use SplObjectStorage;
use SplObserver;

class CustomerAccount implements SplSubject
{
    private SplObjectStorage $observers;
    public function __construct()
    {
        $this->observers = new SplObjectStorage();
    }

    public function attach(SplObserver $observer): void
    {
        $this->observers->attach($observer);
    }

    public function detach(SplObserver $observer): void
    {
        $this->observers->detach($observer);
    }

    public function notify(): void
    {
        foreach ($this->observers as $observer) {
            $observer->update($this);
        }
    }

    public function cancelSubscription(): void
    {
        // Required code for the actual cancellation
        // ...
```

```
        $this->notify();
    }
}
```

A lot has happened here, so let us go through it bit by bit. The first thing notable is that the class makes use of the SplSubject and SplObserver interfaces, as well as the SplObjectStorage class. Since the CustomerAccount class implements the SplSubject interface, it has to provide the attach, detach, and notify methods.

We also use the constructor to initialize the $observers property as SplObjectStorage, which will store all observers of the CustomerAccount class. Luckily, the SPL provides the implementation of this storage already, so we do not need to do it.

> **Standard PHP Library**
>
> We talked about the **Standard PHP Library** (SPL) in *Chapter 3, Code Quality Metrics* already. The fact that the SPL includes these entities shows the importance of the Observer pattern as well as the usefulness of this library.

The attach and detach methods are required by the SplSubject interface. They are used for adding or removing observers. Their implementation is easy – we just need to forward the SplObserver object to SplObjectStorage in both cases, which takes over the necessary work for us.

The notify method has to call the update method on all SplObserver objects that are stored in SplObjectStorage. This is as simple as using a foreach loop to iterate over all SplObserver entries and call their update method, passing over a reference to the subject using $this.

The following code shows what such an observer could look like:

```
class CustomerAccountObserver implements SplObserver
{
    public function __construct(
        private MailService $mailService
    ) {}

    public function update(CustomerAccount|SplSubject
      $splSubject): void
    {
        $this->mailService->sendEmail(
            'sales@example.com',
            'Account ' . $splSubject->id . ' has cancelled
              the subscription'
```

```
        );
    }
}
```

The observer, not surprisingly, implements the `SplObserver` interface. The only required method is `update`, which gets called from the subject in the `notify` method. Since the interface requires the `$splSubject` parameter to implement the `SplSubject` interface, we have to use that parameter type hint. It would lead to a PHP error otherwise.

Since we know that, in this case, the object is actually a `CustomerAccount` object, we can add this type hint as well. This will enable our IDE to help us with the proper code completion; it is not required to add it though.

As you can see, all the logic regarding the email sending has now moved into `CustomerAccountObserver`. In other words, we successfully eliminated the tight coupling between `CustomerAccount` and `MailService`.

The last thing we need to do is to attach `CustomerAccountObserver`:

```
$mailService = new MailService();
$observer = new CustomerAccountObserver($mailService);
$customerAccount = new CustomerAccount();
$customerAccount->attach($observer);
```

Again, this code example is simplified. In a real-world application, all three objects would be instantiated in dedicated factories and brought together by a DI container.

The Observer pattern helps you to decouple objects with a relatively low amount of work. It has a few drawbacks though. The order in which the observers are updated cannot be controlled; thus, you cannot use it to implement functionality where the order is crucial. Second, by decoupling the classes, it is no longer obvious just by looking at the code which observers are attached to it.

To sum up the topic of *design patterns*, we will have a look at those patterns that are still quite common today but have proven to have too significant drawbacks to be recommended. Curtain up for the Anti-patterns!

Anti-patterns

Not every *design pattern* stood the test of time. Everything evolves, and so do software development and PHP. Some patterns that have been successful in the past have been replaced by newer and/or better versions.

What was once the standard approach to solving a problem a couple of years ago might not be the right solution anymore. The PHP community keeps learning and improving, but this knowledge is not yet evenly distributed. Thus, to make it more obvious which patterns should be avoided, they are often referred to as Anti-patterns – this clearly sounds like something you would not like to have in your code, right?

What does such an **Anti-Pattern** look like? Let us have a look at the first example.

Singleton

Before DI became increasingly popular in the PHP world, we already had to deal with the problem of how to effectively create instances and how to make them available in the scopes of other classes. The **Singleton** pattern offered a quick and easy solution that usually looked something like this:

```
$instance = Singleton::getInstance();
```

The static getInstance method is surprisingly simple:

```
class Singleton
{
    private static ?Singleton $instance = null;

    public static function getInstance(): Singleton
    {
        if (self::$instance === null) {
            self::$instance = new self();
        }
        return self::$instance;
    }
}
```

If the method gets executed, it is checked whether an instance of the class has already been created. If yes, it will be returned; if not, it will be created beforehand. This approach is also called **lazy initialization**. Being lazy is a good thing here because it only gets initialized when it is required, so it saves resources.

The method furthermore stores the new instance in the static $instance property. This is remarkable as the approach is only possible because static properties can have a value without requiring a class instance. In other words, we can store the instance of a class in its own class definition. Also, in PHP, all objects are passed as a reference, that is, as a pointer to the object in memory. Both peculiarities help us to make sure that the same instance is always returned.

The Singleton actually is quite elegant; as it also uses a static method, it needs no instance of the `Singleton` class as well. This way, it can literally be executed everywhere in your code, without any further preparation.

The ease of use is one of the main reasons why Singleton eventually became an Anti-Pattern, since it leads to **scope creep**. We explained this problem in the section about DI.

Another problem is testability: it is very hard to replace the instance with a mock object, so writing unit tests for code that uses the Singleton pattern became much more complex.

Nowadays, you should use DI together with a DI container. It is not as easy to use as the Singleton, but that in turn helps us to think twice before we use another dependency in a class.

However, it does not mean that the Singleton pattern must not be used at all. There might be valid reasons to implement it, or at least to keep it in a legacy project. Just be aware of the risks.

Service locator

The second pattern that could be considered problematic is **Service Locator**:

```
class ServiceLocatorExample
{
    public function __construct(
        private ServiceLocator $serviceLocator
    ) {}

    public function fooBar(): void
    {
        $someService = $this->serviceLocator
            ->get(SomeService::class);
        $someService->doSomething();
    }
}
```

In this example class, we inject `ServiceLocator` during the construction time of the object. It is then used throughout the class to fetch the required dependencies. In these regards, DI and the Service Locator are both implementations of the **dependency inversion** principle (the "D" in **SOLID**): they move the control about their dependencies out of the class scope, helping us to achieve a loosely coupled architecture.

But, if we only need to inject one dependency instead of many, is that not a great idea? Well, the drawback of the Service Locator pattern is that it hides the dependencies of the class behind the `ServiceLocator` instance. While with DI you can clearly see which dependencies are used by looking at the constructor, you cannot do that when injecting only `ServiceLocator`.

Unlike DI, it does not force us to question which dependencies should be used in a class as, for larger classes, you can quickly lose the overview of which dependencies are used in a class. This is basically one of the main drawbacks we identified for the Singleton pattern.

Yet again, we do not want to be dogmatic when it comes to the use of the *Service Locator* pattern. There might be situations where it is appropriate to use it – just handle it with care.

Summary

In this chapter, we discussed the importance of standards and guidelines. Coding standards help you to align with fellow developers on how the code should be formatted, and you learned about existing standards worth adopting.

Coding guidelines help your team to align on how to write software. Although these guidelines are highly individual for each team, we provided you with a good set of examples and best practices to build your team's guidelines. With code reviews, you also know how to keep the quality up.

Finally, we introduced you to the world of design patterns. We are confident that knowing at least a good part of these patterns will help you to design and write high-quality code together with your team members. There is much more to explore on this topic, and you will find links to some great sources at the end of this chapter.

This almost ends our exciting journey through the many aspects of clean code in PHP. We are sure you now want to use all your new knowledge in your daily work as soon as possible. Yet, before you do, bear with us for the last chapter, when we talk about the importance of documentation.

Further reading

- `https://google.github.io/eng-practices/review` provides you with more information about the *code review* process at Google

- Useful sources about *design patterns* in PHP:

  - `https://refactoring.guru/design-patterns/adapter/php/example`

  - `https://sourcemaking.com/design_patterns/adapter/php`

  - `https://designpatternsphp.readthedocs.io/en/latest/README.html`

13

Creating Effective Documentation

Many developers consider documentation as a burden rather than a meaningful activity. This is comprehensible since often enough, the documentation is not updated anymore after it has been written. Soon, it is full of wrong statements and outdated information, which is indeed something that nobody wants.

We are convinced that documentation is too important to abandon it. If done right, it will be a valuable addition and an important building block for writing clean code, especially when Working in a Team.

Therefore, in the last chapter of this book, we want to give you some ideas about how to write documentation that is practical and maintainable.

We are going to cover the following main topics in this chapter:

- Why documentation matters
- Creating documentation
- Inline documentation

Technical requirements

For this chapter, there are no additional technical requirements. All code samples can be found in our GitHub repository: `https://github.com/PacktPublishing/Clean-Code-in-PHP`.

Why documentation matters

Welcome to the last chapter of this book. You have come a long way, and before you put this book down, for the time being, we want to draw your attention to the often-neglected topic of creating documentation. Let us convince you on the following pages that documentation does not necessarily have to be tiring and annoying, but instead has valuable benefits.

Why documentation is important

Why should we actually create any documentation? Is our code, or our tests, not enough documentation already? There is some truth in these thoughts, and we will discuss this topic further in this section. Yet over the years, countless developers have never stopped creating countless documents, so there must be something about it.

We create documentation because we can make it easier for other people to work with our software. It is about context, which cannot be easily extracted from reading the code of a couple of classes. Documentation is often not only about the *what* or *how*, but also about the *why*.

Knowing the motivation or the external factors that lead to a decision is crucial to understanding and accepting why a project was built in a certain way. For example, you might complain about your former colleague who has implemented a brittle, cronjob-triggered download of **comma-separated values** (**CSV**) files from an external **File Transfer Protocol** (**FTP**) server, unless you learn from the documentation that the customer was simply unable to provide a **REpresentational State Transfer** (**REST**) **application programming interface** (**API**) endpoint to deliver the data before the project deadline.

A new colleague who starts on your project will surely be happy to have at least some documentation to read, to not need to ask (and probably disturb) other developers for every little question. And let us not forget our future selves, who have not touched that project for many months and must fix a critical bug now. If only we knew what our past selves did back in the day... and why.

If you create **open source software** (**OSS**), then documentation is also important. If you need to evaluate several third-party packages to decide which one to use in a project, it is more likely that a package will be considered if it has good documentation. Wouldn't it be a shame if you invested countless hours in a tool, but nobody uses it because it has no or no good documentation?

Lastly, if you do software development for a living, you should consider it part of the duty of a professional developer to write documentation. This is what you get paid for.

Developer documentation

When we think of documentation, it is usually *user documentation* that comes to our minds first: long, hard-to-read, and boring texts about how to use every single feature of a software product, such as—for example—a word processor. Of course, this documentation exists for a good reason, but that should not interest us in the context of this book, as it is usually not written by (and for) developers.

Software documentation is an extensive field, and as such, cannot be covered in total in this chapter. We rather want to focus on documentation that supports you in the development process and enables you to write *clean code*, as described in the following, non-exhaustive list:

- **Administration and configuration guides**: Besides the obvious need to describe how to install and configure the software, make sure to include a section about *code quality*. This should contain information about which tools are used locally, and how they are configured.

- **System architecture documentation**: As soon as your project becomes big enough that the basic server setup (usually a web server and database on one physical machine) becomes a bottleneck, and you start scaling it, you should think about documenting your infrastructure as well. Eventually, this will save you and others a lot of time searching for the correct **Uniform Resource Locators** (**URLs**) or server accesses, especially in critical situations. It might be a good place to add information about the **continuous integration** (**CI**) pipeline as well.

- **Software architecture documentation**: How is your software built internally? Does it use *events* to communicate between the modules? Are there any queues that should be used? Questions such as these should be answered in the *software architecture documentation*. This makes it easier for other developers to follow the principles.

- **Coding guidelines**: In addition to the software architecture documentation, *coding guidelines* offer advice on how to write the code. We discussed this topic in depth in *Chapter 12, Working in a Team*.

- **API documentation**: If your **PHP: Hypertext Preprocessor** (**PHP**) application has an API that is used by other developers or even customers, you need to provide a good overview of the API functionality. This makes theirs and your life easier, as you will have fewer interruptions from people who want to know how the API works. You can also give good examples of how to build additional API endpoints.

In the next section, we would like to have a closer look at how writing these types of documentation can be made easier.

Creating documentation

Documentation can be written in many ways. There is no one correct approach, and it is often predetermined by the tools already in use, such as the repository service or the company wiki. Still, there are a few tips and tricks that will help you to write and maintain it, and we want to introduce you to these in this section.

Text documents

Let us first focus on the typical, manually written text documents. The classic approach is to set up a wiki, as these have the great advantage that they can be accessed and used even by less-technical people. This makes them a great choice for companies. Modern wikis, either self-hosted or **software-as-a-service** (**SaaS**), offer a lot of reassurance and useful features such as inline comments or versioning. They also can connect with many external tools, such as ticket systems.

Another option is to keep the documentation close to your code by adding it to the repository—for example, within a subfolder. This is a valid approach as well, especially for smaller teams or open source projects. You should not use bloated formats such as **Word** or **Portable Document Format** (**PDF**), though, and rather focus on text-based formats such as **Markdown**. They are many times smaller, and changes to them are easy to track through the version control history.

The crux with manually written documentation is to keep it up to date. Text files or wikis are patient and do not forget, and over time, many pages of documentation virtually pile up in their storage. It gets problematic when it is unclear which documentation is correct and which is outdated. Once in doubt, it is not trustworthy at all anymore.

The only way to address this problem is to set up a process that makes sure that the documents get updated. In the previous chapter, we already introduced a possible way: *code reviews* in combination with a **Definition of Done** (**DoD**). This makes sure that, whenever we are about to add some new or changed code to our code base, we get reminded by a checklist to update the documentation, if necessary.

In particular, system and software architecture are documented using diagrams. Therefore, in the next section, we want to show you how to effectively create these.

Diagrams

A good diagram is usually much more telling than a long text. There are many free-to-use *diagramming tools* available, and you can choose to either manually draw the diagrams or generate them from text definitions.

Drawing diagrams manually

The classical way of creating a diagram is by using a diagramming tool that allows you to draw it manually. These tools are specifically designed to assist you in the creation process—for example, by offering templates and icon sets, or by maintaining the connecting arrows between objects if they are moved.

A versatile tool that we want to present to you in this chapter is *diagrams.net* (`https://www.diagrams.net`). In fact, we also used it to create illustrations for this book. It offers a library of elements that, for example, can be used to create diagrams such as **Unified Modeling Language** (**UML**) diagrams and *flow charts*. It also offers icons for the most popular cloud providers, such as **Google Cloud Platform** (**GCP**), **Amazon Web Services** (**AWS**), and Microsoft Azure.

If you intend to use it, we recommend saving your diagrams as **Scalable Vector Graphics** (**SVG**). SVG is based on **Extensible Markup Language** (**XML**), and although XML is quite verbose, it still consumes less disk space than graphic formats such as **Portable Network Graphics** (**PNG**).

More importantly, it can be loaded and amended in the editor repeatedly, so you do not have to start over again every time your system changes. Most **integrated development environments** (IDEs) and all browsers will display SVG files as graphical images that can even be scaled indefinitely, and if necessary, they can easily be exported into the most popular image formats.

Generating diagrams from definitions

Not everybody likes to use fiddly editors to draw diagrams, though. Luckily, there is a variety of *diagramming tools* that can generate diagrams from definitions. To demonstrate how they work in general, we chose *Mermaid.js* (`https://mermaid-js.github.io`) as an example. It is written in *JavaScript* and utilizes a Markdown-inspired language to define the diagrams.

Before we check out the advantages of this approach, let us first have a look at a simple example of a flow chart:

```
graph LR
    A{Do you know how to write great PHP code?} --> B[No]
    A --> C[Yes]
    C --> E(Awesome!)
    B --> D{Did you read Clean Code in PHP?} --> F[No]
    D --> G[Yes]
    G --> H(Please read it again)
    F --> I(Please read it)
```

The preceding code would render a diagram, as shown here:

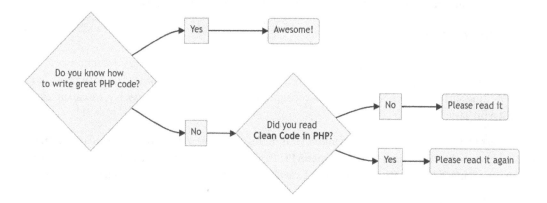

Figure 13.1: Mermaid diagram example

Diagram generation tools help you to create several diagram types, such as *sequence diagrams, Gantt charts*, or even the well-known *pie charts*. You do not have to think about how to style them or how they are arranged. The main work is taken over by the diagramming tool. Of course, Mermaid.js offers many ways to affect the appearance of generated diagrams.

Since the diagram definitions are simple text blocks, they can be added to the code repository. Changes to them are comfortably traceable through the version history. Mermaid diagrams integrate especially well in Markdown documents since the most popular IDEs can display these diagrams directly in the document through additional extensions.

Lastly, if you just want to play around with the possibilities of Mermaid, you can use the *Mermaid Live Editor* (`https://mermaid.live`) to better understand how it works.

> **Mermaid alternatives**
>
> Other noteworthy diagramming tools are *PlantUML* (`https://plantuml.com`), which offers even more practical diagram types to document software architecture, and *Diagrams* (`https://diagrams.mingrammer.com`), which is strong in documenting system architecture.

Documentation generators

Probably the best documentation is one we do not need to create ourselves and is still as useful as human written content. Unfortunately, this will continue to be a dream for now, although we do not know where **machine learning** (**ML**) will take us in the future.

Right now, we can already use tools to create documentation from our code. At least, we can use them to aggregate information that is spread across the many classes of our projects.

API documentation

In this section, we will show you how to create documentation from code using an example of API documentation. If your application provides an API, it is fundamental to have up-to-date documentation for it. Writing such documentation is a time-consuming and error-prone process, but we can at least make it a bit easier.

There exist many approaches to documenting APIs. In this book, we will introduce you to one format that has become more and more popular: **OpenAPI**. This format, formerly known as **Swagger**, describes all aspects of an API in a **YAML Ain't Markup Language** (**YAML**) document, which could look like this:

```
openapi: 3.0.0
info:
  title: 'Product API'
  version: '0.1'
```

```
paths:
  /product:
    get:
      operationId: getProductsUsingAnnotations
      parameters:
        -
          name: limit
          in: query
          description: 'How many products to return'
          required: false
          schema:
            type: integer
      responses:
        '200':
          description: 'Returns the product data'
```

This might be a bit too much information at first glance. Do not worry, though—it is not that complicated. In a nutshell, the preceding YAML describes the Product API in its version 0.1, which offers one endpoint, /product. This endpoint can be called using the **Hypertext Transfer Protocol (HTTP)** verb GET and accepts the optional parameter limit, which is of type integer and must be written in the URL query (for example, /product?limit=50). If all goes well, the endpoint will return with the HTTP code 200.

OpenAPI documentation

The OpenAPI format is quite extensive, so we cannot cover it in this book. If you are interested in learning more about it, please look at the official documentation: https://oai.github.io/Documentation.

As a welcome benefit, IDEs such as *PhpStorm* by default support you in writing these YAML files by checking the validity of the schema. If you, for example, wrote operation instead of operationId, the IDE would highlight the wrong usage.

You can either write a YAML file manually or have it generated. We want to have a closer look at the latter use case. To achieve this, we need the help of a *Composer* package called swagger-php (https://github.com/zircote/swagger-php). Please refer to the package documentation on how to install it.

Of course, the package cannot magically create documentation out of nothing. Instead, swagger-php parses meta information that is written directly in the PHP code, either as *DocBlock annotations* or, with PHP 8.1, as *attributes*. In other words, we need to make sure that the meta information is already there before we can generate the YAML file.

What does this information look like? Let us have a look at the first example, using *annotations*:

```php
/**
 * @OA\Info(
 *      title="Product API",
 *      version="0.1"
 * )
 */
class ProductController
{
    /**
     * @OA\Get(
     *      path="/product",
     *      operationId="getProducts",
     *      @OA\Parameter(
     *          name="limit",
     *          in="query",
     *          description="How many products to return",
     *          required=false,
     *          @OA\Schema(
     *              type="integer"
     *          )
     *      ),
     *      @OA\Response(
     *          response="200",
     *          description="Returns the product data"
     *      )
     * )
     */
    public function getProducts(): array
    {
        // ...
    }
}
```

Based on the information inside the DocBlocks, swagger-php will return the documentation of our API as a YAML file that will look exactly like the preceding example. But why should we use swagger-php when we could write the YAML directly?

Indeed, not everybody wants to have big blocks of documentation within the code, and depending on the level of detail you want to document, they can become much bigger than in our previous example. If you think of an API with many endpoints that are scattered across various controllers in your code, though, you might already realize the benefits: all required meta information is stored close to the code, so if changes are made on the endpoint, it is much easier for the developer to simply amend the DocBlock annotations than to do these changes in some additional document or wiki. Since the comments are part of the code, the changes are also already under version control. In the end, the decision to use swagger-php is up to you or the team.

In the *Inline documentation* section of this chapter, we will discuss why DocBlocks are not the best place to store meta information. Since PHP 8.0, we luckily have a better place for them—namely, attributes, which we already talked about in *Chapter 6, PHP is Evolving- Deprecations and Revolutions*.

Before we discuss why they are the better option, let us have a look here at how our endpoint would be documented by using attributes:

```php
use OpenApi\Attributes as OAT;

#[OAT\Info(
    version: '0.1',
    title: 'Product API',
)]
class ProductController
{
    #[OAT\Get(
        path: '/v2/product',
        operationId: 'getProducts',
        parameters: [
            new OAT\Parameter(
                name: 'limit',
                description: 'How many products to return',
                in: 'query',
                required: false,
                schema: new OAT\Schema(
                    type: 'integer'
                ),
```

```
            ),
        ],
        responses: [
            new OAT\Response(
                response: 200,
                description: 'Returns the product data',
            ),
        ]
    )]
    public function getProducts(): array
    {
        // ...
    }
}
```

Admittedly, the attribute syntax might look a bit unfamiliar. We recommend using attributes instead of annotations, however, because they come with convenient advantages. Firstly, they are real code; they get parsed by the PHP interpreter, and your IDE will be able to support you in writing them. In the first line of the preceding example, you can see that we need to import the OpenApi\Attributes namespace to make this example work.

Within this namespace, you will find actual classes that are referenced here. The files are located inside the vendor folder of your project. This enables you to use features such as autocompletion, and you will get immediate feedback from your IDE if something is not correct, which makes the writing of such documentation much easier.

As the last step, you need to generate a YAML file from the code. This step can, of course, be automated in the *CI pipeline*, which we introduced in *Chapter 11, Continuous Integration*. You can find examples of the usage in our Git repository for this book.

You might wonder: what can I do with this API documentation? Surely, it can already function as documentation for other developers, but there is much more to it. You can, for example, import it into HTTP clients such as *Insomnia* or *Postman*. That way, you can immediately start to interact with the API without having to look up the exact schema.

Another use case is to help you write functional tests for your API. There are packages such as *PHP Swagger Test* (https://github.com/byjg/php-swagger-test) or *Spectator* (https://github.com/hotmeteor/spectator) that can assist you in writing tests against the **OpenAPI Specification** (**OAS**), which can also be considered as a *contract*. You can test, for example, if the returned object for a specific HTTP status code equals what is specified in that contract.

Lastly, and probably the most important use case, is to use the OAS specification with *Swagger UI* (`https://github.com/swagger-api/swagger-ui`), which is a visual and interactive documentation of your API.

The following screenshot shows what our example API would look like:

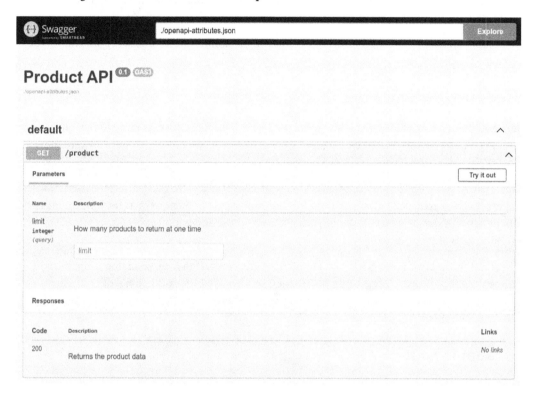

Figure 13.2: Swagger UI

Exploring all possibilities of OpenAPI and Swagger UI would go beyond the scope of our book. We recommend you check out both tools if you want to learn more about them.

OpenAPI alternatives

There are other formats such as **RESTful API Modeling Language (RAML)** (`https://raml.org`) or *API Blueprint* (`https://apiblueprint.org`) that you could use, and we are not opinionated toward any solution.

Inline documentation

A special case is documentation that many of us have done regularly since we started writing software: comments. These are written directly in the code, where the developers can immediately see them, so that seems to be a good place to put documentation. But should comments really be seen as or used for documentation?

In our opinion, comments should generally be avoided. Let us have a look at some arguments on the next pages.

Annotations are no code

Comments are not part of the code. Although it is possible to parse comments through the Reflection API of PHP, they were originally not meant to store meta information. Ideally, your software should still work the same after stripping out all comments.

Today, though, this is often not the case anymore. Frameworks and packages such as **object-relational mappers** (**ORMs**) use DocBlock annotations to store information in them, such as route definitions or relations between database objects. Some *code quality tools* use annotations to control their behavior on certain parts of the code.

PHP cannot throw error messages if your annotations are wrong. If they serve an important purpose, your tests will hopefully catch the bug before they are deployed to production. A better choice is attributes, which are a real language construct. We discussed these in more detail earlier in this chapter when we talked about API documentation.

Unreadable code

Furthermore, as we already discussed in *Chapter 12, Working in a team*, comments are often an indicator of code that is too complex. Instead of explaining your code, you should rather aim to write code that does not need to be commented in the first place.

It can be a fun exercise to compress a whole function into a one-liner—for example, by using some quadruple-nested ternary operators or a frighteningly complicated `if` clause that nobody will understand. You will regret having written it latest at the point when there is a high-priority bug in the production environment, and it is on you to fix it without having the slightest idea anymore what your cryptic masterpiece is supposed to do.

Or, even worse, your new colleague on their first on-call shift has the honor of debugging late at night, when the alerts keep coming in. There are better ways to start a working relationship.

Outdated comments

A comment is quickly written, but also quickly forgotten. Since comments are not parsed by the PHP interpreter, you will not get informed when they are not correct anymore—for example, when the function they are supposed to explain gets rewritten and suddenly serves a different purpose. There is no other way to validate the comments than a developer trying to read and understand their meaning, and comparing it with the actual function code.

At the time of writing, this might not sound like a problem but imagine coming back to a class after a year and finding a comment that you do not understand anymore. Why did you write it in the first place? And if you do not know why, how is anybody else supposed to know?

Outdated comments are wrong information within your code. They are distracting and costly since the developer time does not come for free. Therefore, think twice before you add them.

Useless comments

Try to avoid comments that state the obvious and do not add any further information. The following code snippet is a real-life example of this:

```
// write the string to the log file
file_put_contents($logFileName, $someString)
```

Although it could be considered a nice gesture that the developer took the time to explain the `file_put_contents` function, it does not add value to the code. If you do not know a function, you can look it up. Other than that, it is just an unnecessary line of code you need to scan when reading the code.

It is sometimes not easy to draw a line between useful and useless comments. You could use code reviews to cope with this problem; as discussed in *Chapter 12, Working in a Team*, having somebody else from your team do honest reviews of your code will help to avoid comments such as these.

Wrong or useless DocBlocks

We already discussed DocBlocks and what makes them problematic in *Chapter 12, Working in a Team*, when we introduced coding guidelines. In short, since DocBlocks are basically comments (yet following a certain structure), they can get outdated or simply go wrong quickly if— for example—the parameters of a function call change and necessary changes were not updated in the DocBlock too. Your IDE might throw warnings, but PHP will not.

With the introduction of better type hinting in PHP, many DocBlocks can simply be removed. The redundancy is of no benefit and can rather confuse the reader if the actual code diverges from the annotations.

TODO comments

Comments are not a suitable place to store tasks. You probably know comments such as this:

```
// TODO refactor once new version of service XY is released
```

While some IDEs can assist you in managing your TODO comments, this approach will only work if you are the only one on the project. As soon as you are Working in a Team, using a work management tool such as JIRA, Asana, or even Trello, writing such a comment is simply a way of creating *technical debt*, or, in other words, you are offloading the task into some uncertain day in the future. Somebody else will hopefully fix it one day—most of the time, this will not happen, though.

Instead of a comment, consider creating a task in your work management tool of choice. This way, it is transparent to your colleagues, and it is much easier to plan this work.

When commenting is useful

After discussing what should not be commented, are there any use cases left where comments are useful? Indeed, not so many, but there are some occasions where comments still make sense, such as in the following cases:

- **To avoid confusion**: If you can anticipate that other developers might wonder why you chose that implementation, you should add more context by adding a comment.

- **When implementing complex algorithms**: Even if we try to avoid it, we sometimes have to write code that is hard to understand—for example, if we need to implement a certain algorithm or some unknown business logic. In these cases, a brief comment can be a lifesaver.

- **For reference purposes**: If your code implements some logic that is already explained elsewhere—for example, in a wiki or a ticket—you can add a link to the corresponding source to make it easier for others to find more information about it. This should only be an exception and not the rule.

Please bear in mind that we do not want to be dogmatic. If you feel a comment is needed at some point, write it. It can still be deleted, probably after discussing the topic with another developer in the code review.

Tests as documentation

Developers who write tests often say that these tests also function as documentation. We, too, made the claim in *Chapter 10, Automated Testing* when we talked about the benefits of *automated tests*.

If you do not know what the purpose of a class is, you can at least infer its expected behavior from the tests, because this is precisely what tests do: they make assertions that the code will be tested against. By looking at these assertions, you know what the code is supposed to do.

If the tests fail, you at least know that there is a discrepancy between the assertions and the actual code, and you cannot trust them now. Unless test failures are not generally ignored in your project, you can be sure that someone will fix them soon—or, in other words, the implicit documentation gets updated.

If all the tests pass, you know that you can trust the class implementation—given the tests are well written and do not just test the implementation of a mock object, as discussed in *Chapter 10, Automated Testing*, when we talked about *unit tests*.

Surely, reading and understanding tests are not the easiest form of documentation, but they can be a reliable **source of truth** (**SOT**) if there is no other documentation. They should, however, not be the only type of documentation in your project.

Summary

Writing clean code is not only knowing how to do it yourself but also about making sure that other developers will follow this path too. To be able to do this, they need to know the rules that apply to the project.

In this chapter, we discussed how to create documentation that can help you to achieve this goal. We discussed best practices for manually writing documentation, as well as creating informative and at the same time maintainable diagrams. Lastly, we introduced ways to generate documentation from the code and elaborated on the pros and cons of inline documentation.

Congratulations! You made it to the end of this book. We hope you enjoyed reading it and are now fully motivated to write clean code.

You will probably not succeed with it right away. Strengthening your coding skills is a process that can be frustrating and sometimes even hard to do when working on a commercial project. Try to be patient, and over time you will get better and better.

Reading just one book about clean code is surely not enough. Over the course of this book, we were often only able to merely scratch topics on the surface, and we encourage you to dive deeper into the topics that interest you—and into those that you might not consider interesting in the first place. It requires more studies, an open mind, and the willingness to accept feedback from others to grow your skills.

Yet we are convinced that with this book, we gave you more than a solid starting point for your future journey as a great PHP developer. We would be glad if you think that too.

Further reading

If you want to learn more about Mermaid.js, we recommend the book *The Official Guide to Mermaid. js* by *Knut Sveidqvist* and *Ashish Jain*, published by Packt in 2021.

Index

Packt.com

Subscribe to our online digital library for full access to over 7,000 books and videos, as well as industry leading tools to help you plan your personal development and advance your career. For more information, please visit our website.

Why subscribe?

- Spend less time learning and more time coding with practical eBooks and Videos from over 4,000 industry professionals

- Improve your learning with Skill Plans built especially for you

- Get a free eBook or video every month

- Fully searchable for easy access to vital information

- Copy and paste, print, and bookmark content

Did you know that Packt offers eBook versions of every book published, with PDF and ePub files available? You can upgrade to the eBook version at packt.com and as a print book customer, you are entitled to a discount on the eBook copy. Get in touch with us at customercare@packtpub.com for more details.

At www.packt.com, you can also read a collection of free technical articles, sign up for a range of free newsletters, and receive exclusive discounts and offers on Packt books and eBooks.

Other Books You May Enjoy

If you enjoyed this book, you may be interested in these other books by Packt:

PHP 8 Programming Tips, Tricks and Best Practices

Doug Bierer

ISBN: 978-1-80107-187-1

- Gain a comprehensive understanding of the new PHP 8 object-oriented features
- Discover new PHP 8 procedural programming enhancements
- Understand improvements in error handling in PHP 8
- Identify potential backward compatibility issues
- Avoid traps due to changes in PHP extensions
- Find out which features have been deprecated and/or removed in PHP 8
- Become well-versed with programming best practices enforced by PHP 8

PHP Web Development with Laminas

Flávio Gomes da Silva Lisboa

ISBN: 978-1-80324-536-2

- Discover how object-relational mapping is implemented with laminas-db
- Understand behavior-driven development concepts to sharpen your skills
- Create lean controllers and flexible views
- Build complete models and reusable components
- Explore the Eclipse platform for developing with Laminas
- Find out how easy it is to generate HTML forms with laminas-form
- Practice test-driven development to write precise programs

Packt is searching for authors like you

If you're interested in becoming an author for Packt, please visit `authors.packtpub.com` and apply today. We have worked with thousands of developers and tech professionals, just like you, to help them share their insight with the global tech community. You can make a general application, apply for a specific hot topic that we are recruiting an author for, or submit your own idea.

Hi!

We're Alexandre Daubois and Carsten Windler, authors of *Clean Code in PHP*. We really hope you enjoyed reading this book and found it useful for increasing your productivity and efficiency.

It would really help us (and other potential readers!) if you could leave a review on Amazon sharing your thoughts on this book.

Go to the link below or scan the QR code to leave your review:

`https://packt.link/r/1804613878`

Your review will help us to understand what's worked well in this book, and what could be improved upon for future editions, so it really is appreciated.

Best wishes,

Carsten Windler   Alexandre Daubois

www.ingramcontent.com/pod-product-compliance
Lightning Source LLC
Chambersburg PA
CBHW060535060326
40690CB00017B/3495